Ecobiology of a Hemichordate Worm Saccoglossus sp. From Sundarbans Mangrove complex, West Bengal , India

DR. BHRIGU NATH SINGH

BlueRose ONE.com
Stories Matter
NewDelhi • London

BLUEROSE PUBLISHERS
U.K.

For permissions requests or inquiries regarding this publication, please contact:

BLUEROSE PUBLISHERS
www.BlueRoseONE.com
info@bluerosepublishers.com
+4407342408967

ISBN: 978-93-5741-130-1

Cover design: Daksh
Typesetting: Tanya Raj Upadhyay

First Edition: April 2025

Dedicated to my parents and teachers

ACKNOWLEDGEMENT

My sincere acknowledgements are to the authorities of the ***Susama Devichoudharani Marine Biological Research Institute, Sagar Island, West Bengal*** for the invaluable facilities which I enjoyed during my research period.

I wish to express my deep gratitude and sincere acknowledgement to my teacher ***Dr. Amalesh Choudhury, Professor in Zoology and Head, Department of Marine Science, Calcutta University***, for his supervision, guidance and encouragement, to explore the rigorous vergin part of the Sundarbans mangrove environment and for the preparation of this dissertation.

I also express my sincere gratefulness to the authorities of the department of Zoology, Calcutta University, for allowing me to enjoy the laboratory and library facilities in the department. I express my deep regards for all the reverend Professors of the department of Zoology.

Most gratefully, I acknowledge the financial assistance rendered by the Department of Science & Technology and Department of Environment, Government of India, to make this research programme, a grand success.

I am also grateful to Dr. B.K. Tikader, ExDirector, Zoological Survey of India for providing me the Z.S.1. library facilities and constant encouragements.

I express my sincere acknowledgements to Dr. A. Ghosh, Department of Chemistry, Bose Institute and Dr. (Miss) S. Misra, Lecturer, Department of Marine Science, Calcutta University for rendering invaluable assistance in conducting biochemical analysis of materials in their laboratories and for constructive suggestions throughout the research period.

I am grateful to Dr. R. Sur, Lecturer, Department of Zoology, Calcutta University for providing me laboratory facilities and assistance for analysing gastrointestinalenzymes of the samples, and Prof. S. Gupta, Department of Agriculture, Calcutta University for helping me to analyse various soil parameters..

I am thankful to Dr. T.K. Jana and Dr. S.K. Sarkar, Lecturer in Marine Science, Department of Marine Science, Calcutta University and my colleagues in the Department of Marine Science, Calcutta University for their kind help and valuable suggestions in various ways for the final preparation of this thesis.

I wish to record my deep sense of gratitude to Mrs. Dipti Choudhury. wife of my respected teacher Prof. Amalesh Choudhury for her constant encouragement and appreciation throughout the strenuous period of my research work.

My heartfelt thanks are due to the typists Sri Samar Ghosh and Sri Sajal Kumar Das, and the artistphotographer Sri Nitai Chandra Das for rendering invaluable help for preparation of this thesis. My best loves are left for my associates Sri Rishilal Das, Sri Jalad Kumar Gayen, Sri Asoke Kumar Maity and all the crew members of the research viessel R.V. Sagarputra who helped a lot in the rigorous environment of Sundarbans.

Lastly, I must express my gratefulness to my wife for the trouble she took during all these tough years and to my little loving son for making himself used to the virtual absence of paternal care.

PLATE 1

Saccoglossus sundarbanensis sp. nov.

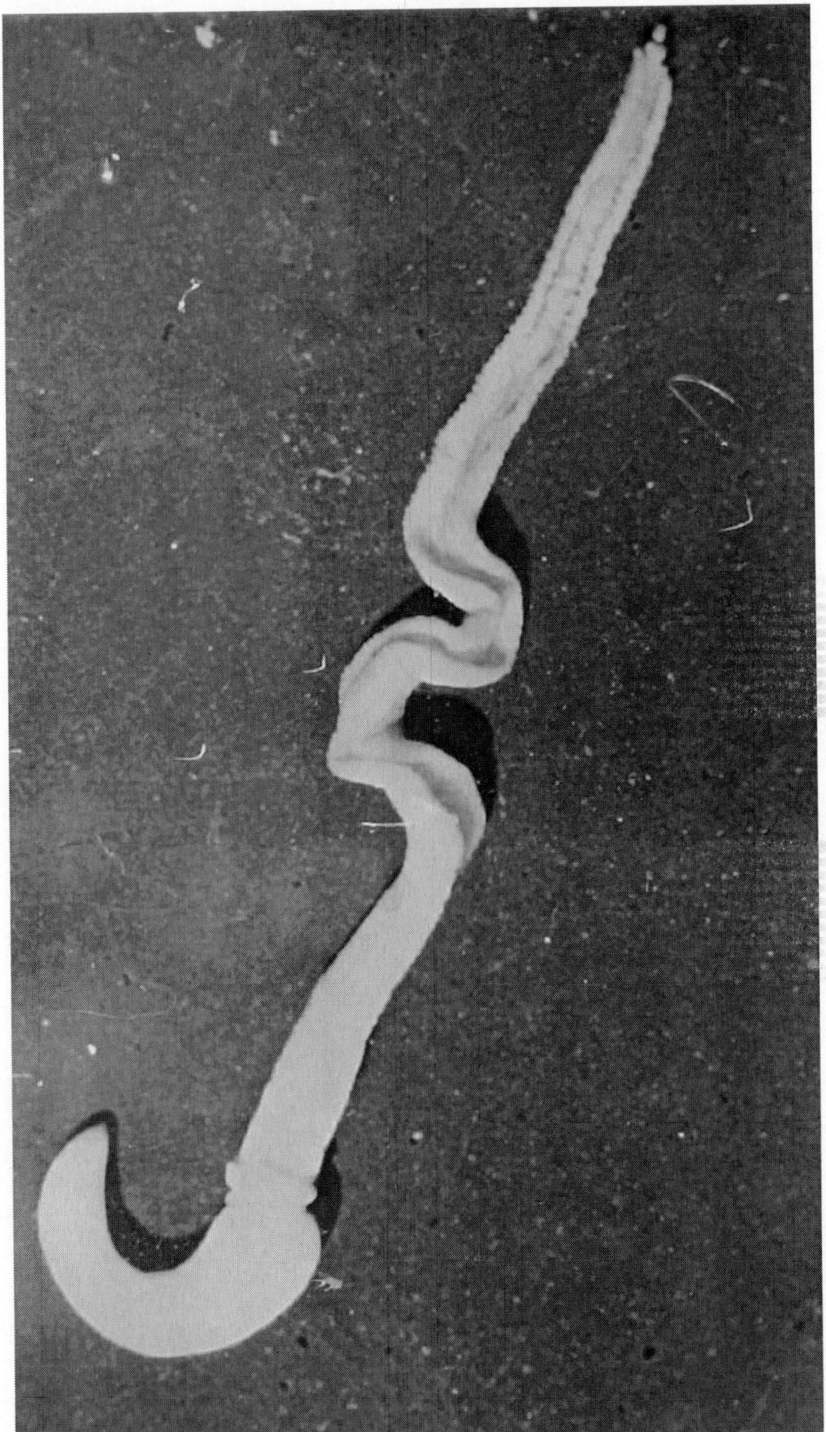

FOREWORD

Mob. : 91-9415757747
Email: omkaar55@hotmail.com
omkar.lkouniv@gmail.com
Resi. : 14/494, Vikas Nagar,
Lucknow - 226 022

Department of Zoology
University of Lucknow, Lucknow-226007 (INDIA)

Dr. Omkar
FNASc, FZSI, FESI, FAEB, FSBA, FAB, FBRS, FAS(AW)
Professor & Head
Receipient of Saraswati Samman

March 25, 2025

Foreword

I have the great pleasure in writing the foreword to the book "**Ecobiology of a Hemichordate worm *Saccoglossus* sp. from Sunderbans Mangrove Complex, West Bengal, India**".

This book is an outcome of sole and dedicated contribution of **Dr. Bhrigunath Singh** on this rare topic among the Hemichordates. The specimens collected by him during the course of his Ph.D. from several islands of Sundarban areas have been named as *Saccoglossus* sp. nov. He has further made studies on different morphological and anatomical features of the animal, like proboscis skeleton, glomerulus, ciliary organ, coelome, muscles, gonads, nervous system, digestive system, respiratory system, blood vascular system, excretory system, reproductive system, etc. In addition, Dr. Singh has also described various species of the genus *Saccoglossus* which will be quite helpful for teachers and researchers to have an overview of this largely neglected group of Zoology. This publication also contains information pertaining to the burrowing and movement, feeding and digestion, excretion, respiration and osmoregulation, breeding behaviour and biology; and macrobenthic fauna associated to *Saccoglossus* sp. Thus, in my opinion, this publication presents a holistic approach on this newly described specimen, *Saccoglossus*.

This publication will be quite interesting in reading and useful in gathering information on this rare group of animals not only to those interested in the subject but also to the generalists keen in acquiring knowledge in various groups of animals.

I express my best wishes for the success of this publication and congratulate the author for his untiring and sustained efforts.

25.03.2025

(Prof. Omkar)

PREFACE

Deltaic Sundarbans of the Ganga river basin is located in the HooghlyMatla estuarine complex on the eastern seaboard of the Indian subcontinent and supports the world's most exciting and magnificent mangrove ecosystem. This ecosystem inspite of supporting world's most graceful game, also supports a luxuriant flora and fauna, micro, meioand macro, in its hundreds of tributaries, salt marshes, and sand and mudflats, in the form of plankton, nehton and benthos.

A mini marine laboratory, the ***Susama Devichoudhurani Marine Biological Research Institute***, unique of its kind in this environment, has been established in 1969 at Sagar Island, a typical Gangetic delta, situated in the mouth of the river Hooghly. This laboratory has so kindly shouldered the responsibility of providing facilities to the enthusiastic researchers and scientists fascinated in marine and estuarine science, to explore the alluring vista of this unique ecosystem.

The present author gratefully availed this opportunity which rendered possible the preparation of this dissertation on the ecobiology of a hemichordate worm. As no substantial work on this important area has been dealt with in the eastern sector of the world so far, and the Indian subcontinent and South East Asia at large, the author was tempted to undertake this fascinating problem on the rare hemichordate material inhabiting the mud floors of the Sundarbans mangrove swamps.

This investigation forms a small part of the multiyear and multi-tier integrated Mangrove Ecosystem Research Programme in the deltaic Sundarbans which has been launched by the Department of Marine Science and Department of Zoology, Calcutta University, in the year 1980, with the partial financial assistance of the Department of

Science and Technology and the Department of Environment, Government of India.

Different islands of Sundarbans have been chosen as the venue to study this hemichordate worm, its detail morphology, ecology and biology in relation to the prevailing ecological parameters. It is being. hoped that this study venture in the rigorous environment of the estuarine complex may pave the way and accelerate fresh investigation on the many other macrobenthic fauna that abound the intertidal salt marshes of the deltaic Sundarbans of the Indian subcontinent.

For convenience, the dissertation has been presented in two separate Chapters, the First Chapter deals with the morphology and taxonomy of the ***Saccoglossus*** sp. and the Second Chapter deals exclusively with the biology and ecology of the worm.

Plate 2

Susama Devichowdhurani Marine Biological Research Institute, Sagar Island, South 24 Paraganas, West Bengal, India (Field Laboratory)

Table of Contents

CHAPTER 1

INTRODUCTION

Hemichordates, commonly known as acorn worms or tongue worms. are predominantly marine, but a few brackish water forms have been recorded from different estuarine mudflats of the world. Hemichordates are a conspicuous group of animals in the animal kingdom bearing immense evolutionary significance that display a unique combination of chordate as well as nonchordate features. They belong to the deuterostome group of animal assemblage having enterocoelous coeloms and embracing, in particular, the three major groups of animals viz., echinoderms, hemichordates and chordates.

The taxonomic status and phylogenetic relationship of the hemichordates have been somewhat of a puzzle among the zoologists for many years. Many authorities still prefer to shelter them under a subphylum of the phylum Chordata and group them together with the subphyla Tunicata (Urochordata) and Cephalochordata to form a major subdivision Protochordata. This classification is based mainly upon certain characteristics of the hemichordates displaying chordate affinity such as paired gill slits (pharyngotremy) and a supposedly rudimentary notochord'. Most hemichordates also bear a restricted dorsal hollow nerve cord (in the collar region) apart from bearing the ventral epidermal solid nerve cord typical of the invertebrates.

The relationship does not, however, justify the inclusion of the hemichordates in the phylum Chordata. The few similarities just mentioned are far outweighed by important differences. The main among these is the absence in chordates of body and coelomic regionations corresponding to the hemichordates, the latter in turn expressing very faintly the body segmentation (metamerism) and segmentally arranged organ systems of the chordates. To elucidate further this contention Hyman (1959) may aptly be quoted here saying, "Hemichordates are definitely invertebrates in general morphology and in most details of their construction, and these facts are concealed if they are included in the Chordata. Further, they are closure to the echinoderms than they are to the chordates, and if they are to be united with any other phylum would more justifiably be included with echinoderms. The best solution of this situation is to regard them as an independent phylum of invertebrates stemming from an ancestral type that led, on the one hand, to echinoderms and, on the other hand, to hemichordates and chordates".

In many deuterostomes, such as echinoderms and hemichordates, the body is divided, at least in the larval stage, into three regions a protosome. a mesosome and a metasome; and each region contains a coelomic compartment: a protocoel, a mesocoel and a metacoel.

Because of these and other reasons, the hemichordates are placed in a phylum of their own, phylum Hemichordata (Hyman, 1959), which is made up of two main classes **Enteropneusta** and Pterobranchia. Both the classes have tripartite body composed of proboscis, collar and trunk, but the class **Enteropneusta** is adapted for a sluggish and burrowing habit and optioned for suspension or deposit feeding. Planctosphaeroidea, a new class under hemichordata. is being recognised by some authors based on a few pelagic larvae (Damas and Stiasny, 1961: Hadfield and Young, 1983).

Saccoglossus sundarbanensis sp. nov.. an enteropneust hemichordate christened in this dissertation for the first time, has been discovered and reported by Singh and Choudhury in 1984 from the deep mud beds of detritus rich salt marshes of the deltaic Sundarbans (India). This geographical locale (88° to 89°29' E meridians and 21° to 22°30' N latitudes) supports the world's most magnificent mangrove ecosystem sheltered in the famous HooghlyMatla estuarine complex that experience inundation and exposure twice in 24 hours. The new ***Saccoglossus*** species is being claimed to be reported by Singh and Choudhury (1984) for the first time from any mangrove environment. The type material was first discovered from the mudflats of Prentice Island bordered by the Saptamukhi waters, a giant tributary of the river Ganges. Later, the

Saccoglossus specimens have been found and collected by the present author from many other islands and mudflats of Sunderbans mangrove ecosystem viz., Sagar Island, Chandanpiri mudflat of Namkhana range; Lothian island; Dia complex; Bhagabatpur and Bhubaneswari mudflats;and Sajinakhall, Jharkhall and Netidhopani mudflats. all under the Sundarbans Project Tiger, a core area in the virgin mangrove system sheltering the world's largest population of 'Royal Bengal Tiger.

The class **Enteropneusta** Includes three families sheltering some Important genera tabled below:

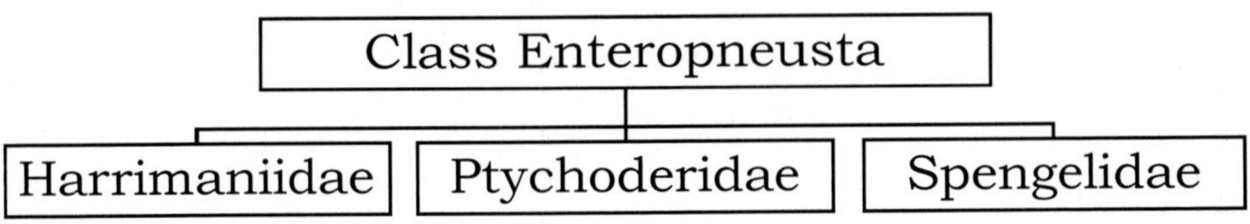

The present part of the dissertation is designed to project specially the genus ***Saccoglossus*** belonging to the family Harimanidae that embraces globally distributed 13 species of which two species, **S. bournel** (Menon, 1904) and **S. madrasensis** Rao, 1955 (?) have been reported from India so far. The author enjoys the privilege to describe and discuss at length the ***Saccoglossus*** sp. Singh and Choudhury, 1984 to determine its taxonomic status and specific identity and hence contribute a new thember to the genus.

*No published document on **S. madrasensis** is available.

HISTORICAL RESUME

Benthos is an Important and essential component of the marine and estuarine biota, but except for the species with obvious commercial value, it received little attention. Earlier investigations on benthos were directed only towards the discovery of species and enumerating them. During the beginning of the present century, interest seemed to be directed more on the benthic ecology with particular reference to benthos as essential component of marine and estuarine food chain system and to understand their structure and function. Through the efforts of Petersen and Boysen Jensen (1911), researchers used quantitative sampling methods and began to describe benthic communities and the environmental factors that might be affecting them. A survey of literature reveals that a lot of works have been carried out on benthic ecology in different marine and estuarine areas of America, Europe and Australia. Notable contributions on this aspect, worth mentioning, are of Gunter (1950, 1956, 1961), Day (1951). Kinne (1963, 1964). Melbourne (1967). Davis (1967). Carriker (1967). Muus (1967). McIntyre (1968, 1969), Holme and McIntyre (1971), Stephen and Edmonds (1972). Tenore (1972). Diener (1973) and Rainer (1981).

Studies on marine and estuarine benthos in our country were initiated by Dr. N. Annandale as early as 1907. More recently several workers have added considerable information, both qualitative and quantitative, to this subject from different parts of the country. Works of Seshappa (1953) on Malabar Coast. Damodaran (1973) on Kerala coast, Deasi and Krishnakutty (1967) on Cochin backwater, Kurian (1967, 1971, 1972) on south east coast and tropical estuary, Parulekar and Wagh (1975) on north eastern Arabian shelf, Parulekar and Dwivedi (1973, 1974) on Goa estuaries, Bhavanarayana (1975) on Kakinada Bay, Harkantra (1975, 1980) on Kali estuary of Karwar and along the west coast of India, Parulekar ***et al.*** (1976, 1980) on

Bombay coast and Goa estuaries. Varshney ***et al.*** (1981) on the benthos of west coast of India.

Other notable contribution are of Fernando ***et al.*** (1981) on Veller estuary, Ansari ***et al.*** (1977) and Harkantra ***et al.*** (1982) on Bay of Bengal. Khan and Natarajan (1975) on Veller estuary, Krishnamurthy (1963) on Adyer estuary.

In the HooghlyMatla estuarine complex. studies on the intertidal benthic ecology are still rather meagre. During the early part on the present century, studies of Kemp (1917) on Matla river and Annandale (1922) on Ganga estuary were the only valuable contributions. Very recently, few quantitative assessments of benthic macrofauna and their ecology in the creeks and intertidal mudflats of Sagar Island and Prentice Island have been documented by Bhunia and Choudhury (1981), Nandi and Choudhury (1983), Choudhury ***et al.*** (1984a, b), Nandi (1986). Chakrabarty ***et al.*** (1986) and Ghosh and Choudhury (1986).

Among the lesser invertebrate deuterostomes. Hemichordates constitute an important representative benthos that claims wide distribution from the deepest ocean floor to the intertidal sand and mud flats. It was in 1825 when F. Eschscholtz first recorded an enteropneust worm burried in sand in shallow water at Odia in the Marshall Islands and named it as Ptychodera flava. This species is now known to be widely spread throughout the tropical IndoPacific waters. The second most Important member of the group was discovered in the Naples region by Delle Chiaje (1829) who gave the animal the name **Balanoglossus** clavigerus, thus creating the most familier generic name in the phylum. A third enteropneust was discovered by Stimpson off the Carolina coast, and described by Girard (1853), who named it Stimpsonia aurantiaca considering it a nemertian; later this was transferred to the genus **Balanoglossus**.

In 1870 Gegenbour proposed the name Enteropneusti for animals of the type of **Balanoglossus**, based on Kowalevsky's discovery (1866) of the presence of gill slits in B. clavigerus.

During the next 30 years several other enteropneust species were discovered and described although the first accurate account of the anatomy was that of Kowalevsky (1866), based on Delle chiaje's species. During that time, it was customary to assign all enteropneust species to the genus **Balanoglossus**, because Eschscholtz's genus Ptychodera had then been over looked. The partition of **Balanoglossus** began with Spengel (1891), who recognised Ptychodera and **Balanoglossus** and proposed in addition **Glandiceps** and **Schizocardium**. In Spengel's famous monograph (1893) all these four genera were considered to contain all the enteropneust species known at that time, comprising 10 previously described species and 9 new ones. one of which, however, was a synonym of Ptychodera flava. Spengel also proposed in the same monograph, the name Dolichoglossus as a subgenus for those species of **Balanoglossus** having an exceptionally long proboscis being unaware that Schimkewitsch [1892) had already erected the genus ***Saccoglossus*** for enteropneusts of this type (Hyman, 1959).

It was not until 1934 that Van der Horst revived the name ***Saccoglossus*** introduced by Schimkewitsch and shifted Dolichoglossus into synonomy with it. In 1901 Spengel recognised two additional enteropneust genera Spengelia (Willey, 1898) and Harrimania (Ritter, 1900) and himself added two more genera, Glossobalanus and Stereobalanus, thereby raising the total number of genera to 12. A total of 63 species of enteropneust worms were included by Dawydoff (1948) in the Traite de zoologie, all distributed over those 12 genera. New species described since then bring the number to about 70.

Hemichordata, as is conceived at present, comprise three existing classes, viz. (1) **Enteropneusta**, (ii) **Pterobranchia** and (iii) **Planctospaeroidea**. The last class was proposed by Van der Horst

(1935) based on the discovery and description of two deepsea pelagic larvae, Planctosphaera pelagica by Spengel (1909).

Johannes Muller (1850) first seen and examined the larva of Enteropneusts at Marseilles and gave it the name tornaria. Later, Krohn (1854). Agassiz (1864) and Muller (1868) observed the tornaria at Messina, New England and Brazil respectively. All these authors recognised it as an echinoderm larva because of its similarities with bipinaria. It was Metschnikoff (1869, 1870) who ascertained that tornaria was the larva of **Enteropneusta**, because at Messina, he was able to follow its development into a young acorn worm. This discovery was soon verified by Agassiz (1872, 1873).

It was Bateson, who in 1885, studied the embryology of some enteropneusts and documented many resemblances of development and structure to the lower chordates, especially Amphioxus, and proposed the name Hemichordata to replace **Enteropneusta**. He suggested and recommended the inclusion of the hemichordates with the tunicates, cephalochordates and vertebrates in the phylum chordata, though this inclusion was being recented regularly by many zoologists and students of **Enteropneusta**, beginning with Spengel (1893).

The inclusion of the hemichordates in the chordates persisted in the classical Zoological works since it was first proposed by Bateson in 1885. Subsequently, however. It was rejected by Van der Horst (1939) and in other great treatises on Zoology, viz. in the Handbuch der Zoologie and Traite de Zoologie. The whole organization of hemichordates differs from that of chordates except for the presence of gill slits and their supporting skeleton in enteropneusts, and the structure once supposed to represent the notochord is now regarded as of other nature (Hyman, 1959).

Hyman (1959) accepted the opinion of Van der Horst (1939) that Hemichordata was the most suitable name for this phylum and she endorsed the opinion of the school in removing the Hemichordata from the Chordata and making them an independent phylum of invertebrates.

Other notable additions to the taxonomic literature for hemichordates may be cited from the reports of Menon (1904) on ***Saccoglossus*** (Dolichoglossus) bournei; Balasubrahmanyan (1959) on the occurrence of **Glandiceps** sp. and ***Saccoglossus*** sp. from Porto Novo, India: BurdonJones and McIntyre (1960) on Stereobalanus, a genus new to the Old World: BurdonJones (1962) on **Balanoglossus** gigas; Ramanujam (1935). and Devanesan and Varadarajan (1940) on the tornaria larva from Indian waters. BurdonJones and Patil (1960) worked for the revision of the genus ***Saccoglossus*** and established 12 species by synonymising S. cambrensis and S. serpentinus with S. ruber.

A number of interesting and very important contributions on the biology, physiology, histochemistry, fine structure, feeding and breeding behaviour, development and morphogenesis in hemichordate worms, have been made by various workers from different parts of the globe. TheHistorical Resume of this dissertation enjoys the privilege to include some of them, viz. Rao (1953, 1954a. b. 1955a. b): Ashworth and Cormier (1967); Barrington (1941), 1965): De Jorge and Peterson (1968a, b); De Jorge ***et al.*** (1967); Krishnan and Govindarajulu (1968): Macha (1969): Azariah (1969, 1975); Peterson and Longhi (1971).

Interest of Zoologists in India in the **Enteropneusta** dates back to 1904 when K. R. Menon made known to the existence of ***Saccoglossus*** (=Dalicoglossus) bournei at the Madras coast and published a monograph on it. Since then seven species of enteropneusts have been reported from the Madras coast by different workers (Gavely. 1922: Ramanujam, 1935: Kuriyan, 1949: Sundara Rao and Ranga Rao, 1949: Rao, 1955b, Balasubrahmanyan. 1959). These are Ptychodera flava from Krusadai island; **S. madrasensis** Rao, 1955; **Glandiceps** coramandelicus Menon, 1904; G. stlasnyi

Rao, 1953; Glossobalanas elongatus spengel; G. minutus (Kowalevsky, 1866) and Ptychodera flava Eschscholtz, 1825.

Very recently, an enteropneust species of the genus ***Saccoglossus*** has been reported by Singh and Choudhury (1984) from the mangrove swamps of the deltaic Sundarbans, West Bengal, India. The present dissertation intends to project at length the detail morphology of the hemichordate worm and to establish its specific identity. Chapter II of this dissertation involves in portraying the ecobiology of the worm in a unique environment.

PHYSIOGRAPHY AND CLIMATE

The geographical location of the present investigation lies in the HooghlyMatla estuarine complex (Fig. 1.) embracing the western sector of deltaic Sundarbans nearly 100 km south of Calcutta. This estuarine complex of the tropics occupies an important place in the world map and can be categorised as a river delta estuary (formed at the mouth of large rivers) which has been advocated by Odum (1971) in addition to Pritchard's (1967) four categories of estuarine complex. There are seven major rivers in the HooghlyMatla estuarine system. most of which are blind in the north except the river Hooghly In the extreme west that receives perennial supply of fresh water draining into the sea through the Hooghly estuary. This estuary is a positive estuary of mixohaline type. It has an approximately funnel shaped wide mouth spread over about 20 km stretch and probably due to the strong scouring action of the stream as well as tidal currents, a greater overall circulation is maintained. The tidal influence is felt in the upper reach of the river upto a distance of 290 km from the sea face. The estuary experiences vertical turbulance to a high degree and is reported to have no appreciable stratification of temperature and salinity. practically throughout its entire length (Bose, 1956).

The deltaic system of the estuary covers several islands of which Prentice island was selected for the present dissertation (Fig. 2).

Prentice, a small island (6 sq. km) is situated in Saptamukhi river complex of the deltaic Sundarbans. The island extends from 88°17'30" E to 88°18'42" E longitude and 21°42'15" N to 21°45'50" N latitude. The height of the island ranges from 3 m to 3.5 m from the mean sea level and the tidal amplitude of the area is over 5 m in average. Thus the area of the island becomes totally inundated in each spring high tides.

This low estuarine deltaic island is divided into two major geomorphological units on the basis of evolution. These units are:

(1) Mangrove swamps and

(2) Marginal mudflats

Important processes like erosion and deposition are initiated by tidal and nontidal residual currents in the Saptamukhi river where the Prentice Island is situated. The northsouth elongation of the island, shaped by the estuarine channel flow of the river Saptamukhi, is located in the predominantly dynamic depositional environment of the Ganges delta. The dynamacity is Induced with the southwest monsoon run off and the cyclonic storm waves in the estuary.

(1) Mangrove swamps

The morphology of the mangrove swamps is characterised by intertidal mudflats and the occurrence of the dense mangrove trees whose branching root systems and twisted trunks and branches provide the requisite resistance to tidal currents. During the equilibrium stage of high tide the suspended silts are laid down and deposited on the mudflat and thus the surface of the swamps is raised very slowly and gradually with the time interval of inundation. The mangrove swamps of the Prentice island are crisscrossed by a network of meandering tidal creek system in which the banks are formed by the combined efforts of roots and the sediment. Tidal flows are carried by the creek systems below the level of the swamp surface. During these low tidal flows, the creeks act as small 'tidal estuaries' and exhibit a range of flow velocities; the sediment transport variationcreekbanks are steep and the depth ranges from 3 m to 4 m.

(2) Marginal mudflats

Prentice island is surrounded by wide marginal mudflats in the Saptamukhi estuary. Mudflats accrete more sediments during the high tide and the erosion takes place in the mid tide. Thus the break of slope is marked on the mudflat along the high tide and the mid tide lines. The slope ranges from 5° to 22° in which the velocity profile can be observed. The fine textured sediments of the mudflat are highly permeable and therefore form the magnificent home of different burrowing organisms. Mudflats are wider (about 300 m) in the south than the north (80 m to 10 m) of Prentice island.

The southern modflats are gradually being wider (5 m to 8 m per year) with the land building activities of mangroves. The mudflats of Prentice island are Indented by several embryo creeks. At their substratum water Interface, the flats, devoid of vegetation, are impregnated with rich detritus.

Seasons

Seasons are pronounced in this estuarine system, each with fourmonth duration and the seasonal variations have been noted in accordance with the tidal regime (Pillay, 1958). The pre-monsoon (March to June) is the dry season with occasionally higher temperature. The monsoon season (July-October) is accompanied by heavy rain fall, and the postmonsoon (November to February) comprises partly the winter season, comparatively with lower temperature and lesser precipitation. With the change of seasons tidal interactions in the Hooghly estuary also change, highest high tide 5.9 m and lowest low tide 0.5 m in the month of August/September.

Wind:

The prevailing wind is from north and northeast from the beginning of October to the middle of March. The months of January and February are relatively calm. It again commences to blow violently from southwest about the middle of March and subsides about September. Storms are common during spring and autumn. Some of these often develop into cyclones of varying intensity usually accompanied by tidal waves and cause much loss of life and damage to property and the forests.

Temperature :

A bimodal temperature oscillation per year was a distinguished feature. Two maxima, one in the pre monsoon months and the other in the monsoon months were observed. The rise of temperature was noticed higher in the pre monsoon months and lower during post monsoon months.

In Hooghly estuary, Bose (1956) reported that the maximum difference between the surface and bottom temperature was found to be 1.5°C. The very low difference in temperature may be attributed to the fact that the bottom of the estuary is uneven, and that the tidal push results in rapid mixing of surface and subsurface waters. The silt particles in the highly turbid water in the estuary might have also helped in the transmission of heat, preventing any marked stratifications (Ellis, 1936).

Precipitation:

The observatory at Sagar Island records maximum value (620 mm) in August in the first year. During the second year the highest value (579.08 mm) was also obtained in August (from 1982 to 1984).

With the change of seasons, tidal interactions in the HooghlyMatla estuarine system also change. Three distinct seasonal variations have been observed in the estuary so far as tidal regime is concerned (Pillay, 1958).

During the rains, from July to October considering the effects of the monsoon first, the flood tides are more or less completely nullified by the freshets and there is a strong predominance of the ebb tides.

The rainy season in Bengal is during the southwest monsoon and unless otherwise stated the word monsoon is used in this dissertation to denote the southwest monsoon only.

From November to February is the second tidal rhythm. Owing to the banking of freshwater in the river surface run off during the monsoon, the strength of the flood tide over the ebb tide is minimum.

On the other hand, during the third rhythm (March to June) the flood tides considerably accentuate over the ebb tide. This brings about change in the salinity gradient of the estuary.

Map of Hooghly – Matla Estuarine Complex

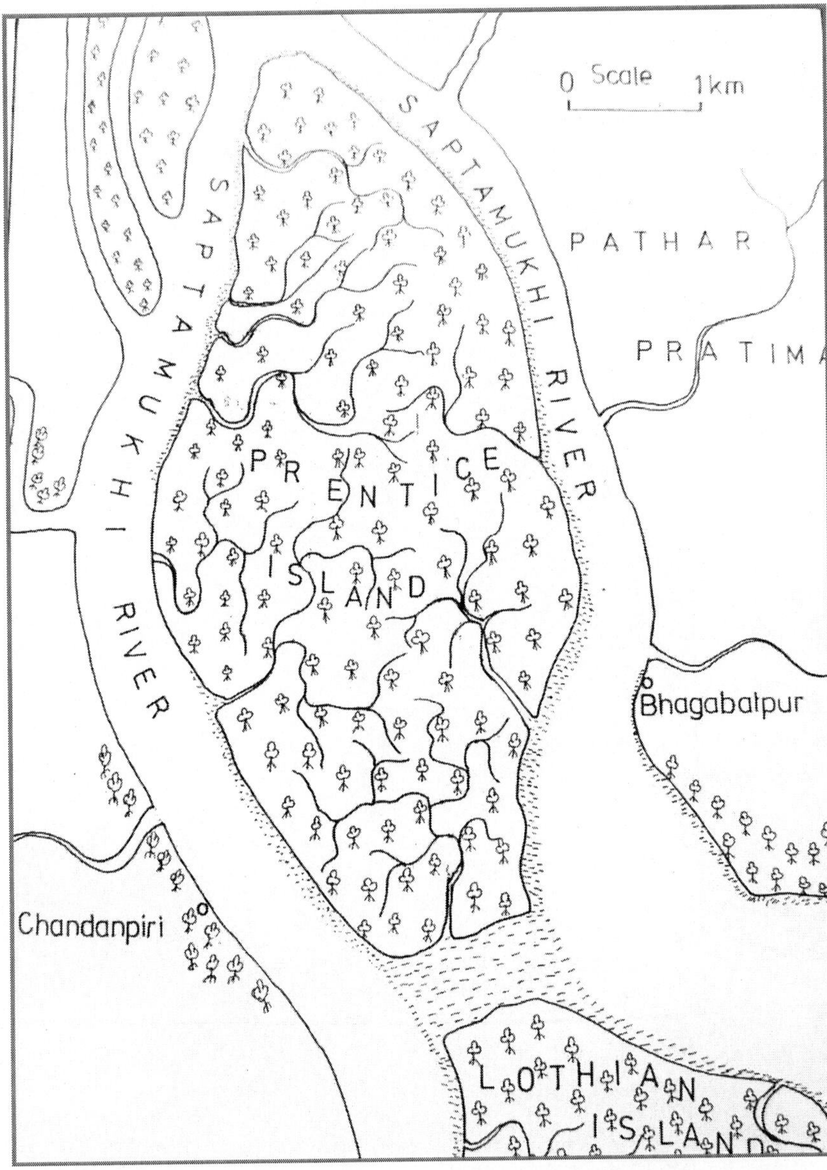

Geographical Locale Of Prentice Island In Sunderbans

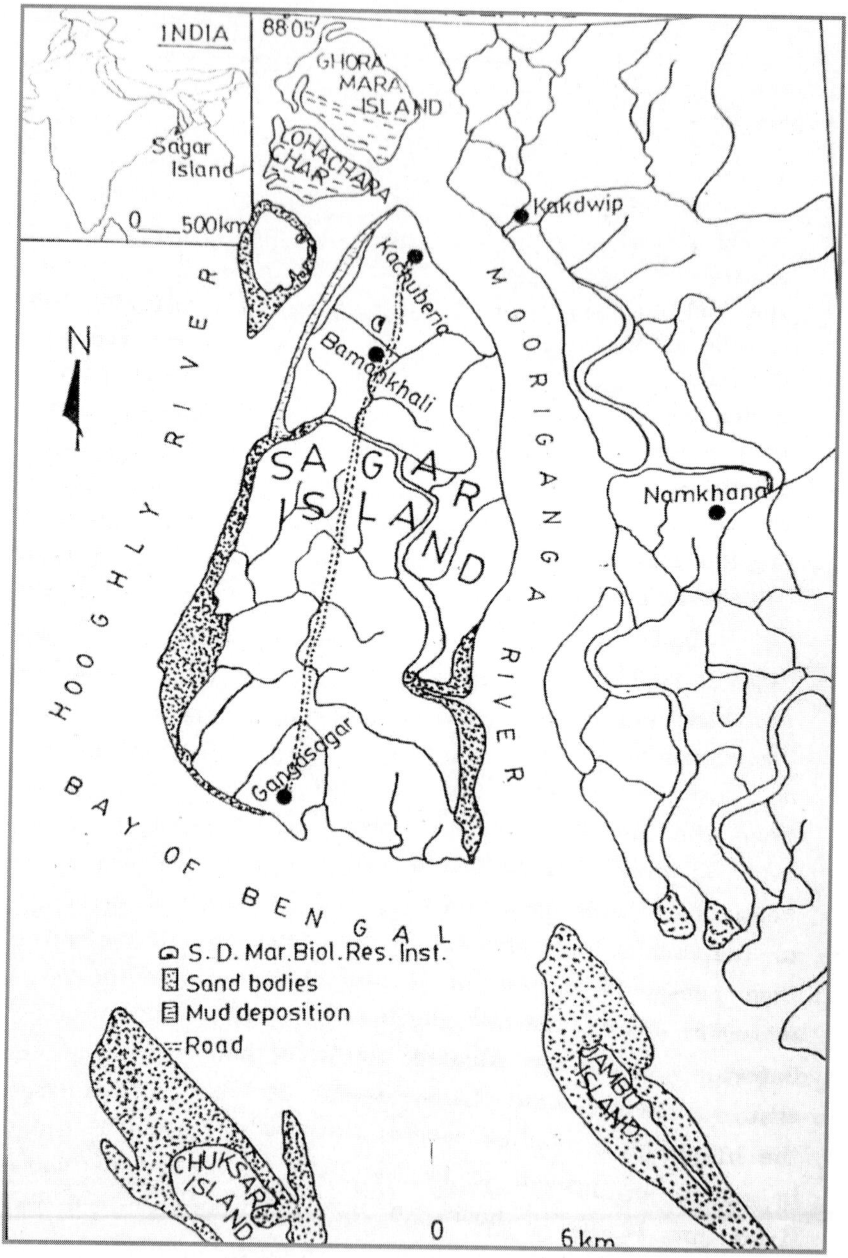

Map of Sagar Island

MATERIAL AND METHODS

The hemichordate worms (***Saccoglossus*** sp.) were collected for the present investigation from the intertidal mudflats of HooghlyMatla estuarine complex (Sagar Island. Prentice. Luthian and Bhagabatpur island clusters, etc.). To collect the hemichordate worms their faecal castings on the exposed mudflats were spotted out during the receding tide and then the soil was removed by hand following the inner zigzag tunnels of the worm. The animals were washed carefully by water of the spot. As the worms are fragile, the rear trunk parts are fragmented during unearthing. The body fragments were kept in a plastic container with sufficient interstitial water and brought to the field laboratory at Sagar Island (S. D. Marine Biological Research Institute). Some of the healthy worms were kept in glass aquarium containing little amount of detritusrich mud and sufficient interstitial water from the same habitat to study the burrowing, feeding and breeding behaviour.

Very weak solution of formalin, dilute alcohol, magnesium chloride ($MgCl_2$) etc., were used for narcotisation of which $MgCl_2$, rendered the best result. The procedure employed here was as follows: the hemichordate worms were kept in a petridish or in an enamel trough and allowed to relax to its fullest extension with just enough water.When the animals were fully expanded. 7.5% $MgCl_2$ solution was added drop by drop at regular intervals. After a few hours, the hemichordate worms lost their peristalsis as well as power of con traction and ceased to respond to the chemical stimuli applied. Narcotised specimens were then preserved either in 4% neutral formalin or in 70% alcohol. Extra formalin or alcohol was added for better and prolong preservation. General anatomy was studied by dissection, while the details of microanatomy was studied from several series of microtome sections. For the histological studies several fixatives were tried, but BouinsHollande in sea water proved to be the best. Several narcotised materials were fixed for 12 to 15 hours in BouinsHollande in sea water for microanatomical studies.

After fixation the materials were washed by passage through a series of changes in 70% alcohol until the picric acid of the fixative was totally removed. The materials were then taken in a paraffin tray and the desired parts of the specimen were cut and placed in different tubes with 70% alcohol, properly labelled, and then passed through a series of alcoholic grades (70%, 90% and absolute alcohol). After dehydration and infiltration, the tissues were embedded in paraffin for microtome sectioning. Both sagital and transverse sections were cut at 6 to 10 um thickness. For general histological study, the sections were stained with haematoxylin and eosin, and also with Malory's triple stain.

Sampling of associated macrobenthos :

Study of the macrobenthos in association with the hemichordate worms in the midlittoral zone of Prentice island was conducted. Collection of macrobenthos was carried out at five different spots taken at random for each station during each season of investigation. The observations were based on monthly collection during low tide, when the mid littoral zone was accessible for proper investigation. Benthic samples were collected with the help of a quadrate having one square meter area. The quadrate was thrust into the mud and carefully removed. without disturbing the contained core of the mud. The hemichordate and other associate fauna retained in the mud were carefully sorted out on a polythene sheet and counted by hand picking method and expressed in no/m^2.

Environmental parameters :

For field ecology study, flood water inundating the midlittoralmudflat, as well as interstitial water, and soil samples were collected from the exact habitat monthly and seasonally for the assessment of various environmental parameters.

The temperature (air, soil, surface water and interstitial water) was recorded with a centigrade thermometer. The salinity of the sample water was estimated by employing 'Mohrknudsen method': the Winkler's method was followed to estimate the dissolved oxygen. For pH, Lovibon's colour comparator disc was used.

Determination of soil texture :

Mechanical analysis was carried out by international pipette method as modified by Robinson (Piper. 1950).

Soil pH was measured by Elico pH meter model WIO with glass electrode using soll water suspension in ratio of 1: 2.5 (Jackson, 1973).

Conductivity was measured by Toshniwal conductivity Bridge type CL 01/01 A with soil water suspension in the ratio 1: 5 (Jackson, 1973).

The available nitrogen of the soil was determined by alkaline $KMnO_4$, method (Subbiah and Asija. 1965). Phosphorus by Olsen's method (Jackson, 1973) and Potassium by Flame photometric method (Jackson, 1973).

The following statistical procedure was used to evaluate the generated data.

Correlation coefficient :

To find out the relationship between different environmental parameters and the relationship between environmental parameters and the population densities of ***Saccoglossus*** sp. the method of Sneoccor and Cochran (1967) has been followed :

Months	Variate (X)	Variate (Y)	XY
1. August, 1982	X_1	Y_1	X_1Y_1
2. September, 1982	X_2	Y_2	X_2Y_2
3. October, 1982	X_3	Y_3	X_3Y_3
|	|	|	|
24. July, 1984	X_{24}	Y_{24}	$X_{24}Y_{24}$
Total	$\sum_{i=1}^{i=24} X_i$	$\sum_{i=1}^{i=24} Y_i$	$\sum_{i=1}^{i=24} X_iY_i$

X = Hemichordate population/burrow temperature/Dissolved O_2/Salinity/pH etc.

Y = Abiotic parameters (Water temperature, burrow temperature, Dissolved O_2, Salinity, pH)

$$\delta X^2 = \frac{\sum_{i=1}^{i=24} X_i^2 - \left[\sum_{i=1}^{i=24} X_i\right]^2 / 24}{24 - 1}$$

$$\delta Y^2 = \frac{\sum_{i=1}^{i=24} Y_i^2 - \left[\sum_{i=1}^{i=24} Y_i\right]^2 / 24}{24 - 1}$$

Correlation between X and Y or

$$rXY = \frac{\sum_{i=1}^{i=24} X_iY_i - \sum_{i=1}^{i=24} X_i - \sum_{i=1}^{i=24} Y_i \Big/ 24 - 1}{\sqrt{\delta X^2} \quad \sqrt{\delta Y^2}}$$

Analysis of Variance:

To find out the variation of <u>Saccoglossus</u> sp. Population over different months and years, two way classifications were used according to the following model:

17

MODEL - I

	Year[1]	Year[1]	Total
Month 1	$X_{1.1}$	$Y_{1.2}$	
Month 2	$X_{2.1}$	$Y_{2.2}$	
Month 3	$X_{3.1}$	$Y_{3.2}$	
Month 12	$X_{12.1}$	$Y_{12.2}$	Grand Total

$$\text{Correction factor (C.F.)} = \sum_{i=1,\ J=1}^{J=12,\ i=12} X_{iJ} \Big/ 24$$

$$\text{Total sum of square (S.S.)} = \sum_{i=1,\ J=1}^{J=2,\ i=12} X_{iJ}^{2} - \text{C.F.}$$

X = Months of first year

Y = Months of second year

J = Column

i = Row

MODEL - II

Month/Year	Y_1	Y_2	Total
M_1	y_1m_1	y_2m_1	$M_1 = \sum_{i=1}^{i=2} y_i m_1$
M_2	y_1m_2	y_2m_2	$M_2 = \sum_{i=1}^{i=2} y_i m_2$
M_{12}	y_1m_{12}	y_2m_{12}	$M_{12} = \sum_{i=1}^{i=2} y_i m_{12}$
Total	$\sum_{j=1}^{i=12} y_1 m_j$	$\sum_{j=1}^{j=12} y_2 m_j$	

$$\text{Month S.S.} = \sum_{j=1}^{j=12} m_j^{\,2} \Big/ 2 \quad - \text{C.F.}$$

$$\text{Year S.S.} = \sum_{i=1}^{i=2} y_i^2 \Big/ 2 - \text{C.F.}$$

$$\text{Error S.S.} = (\text{Total S.S.}) - (\text{Month S.S.} + \text{Year S.S.})$$

y_1m_1 = Population in the month_1 of year_1

y_1m_{12} = Population in the month_{12} of year_1

y_2m_1 = Population in the month_1 of year_2

y_2m_{12} = Population in the month_{12} of year_2

j = Column, i = Row

Anova for characters

Source of variation	Degree of freedom (d.f.)	Sum of square (S.S.)	Mean square (M.S.)	F
Month (M)	11			
Year (Y)	1			
Error (E)	11			
Total	23			

Estimation of lipids and fatty acids :

Extraction of lipids Live adult ***Saccoglossus*** specimens were cut into pieces, and lipid was extracted by homogenising in a waring blender for 2 minutes using methanol and chloroform mixture (2:1, v/v). following the method of Bligh and Dyer (1959). Centrifuged homogenate was decanted and the residue was homogenised for 2 minutes with methanol, chloroform and water mixture (2:1:0.8, v/v/v). Homogenate was again centrifuged, supernatant decanted and the residue was homogenised following the same method and solvent mixture. The decanted supernatants were combined and diluted with excess water. shaken vigorously in a separatory funnel and allowed to settle, till the solvent layers were distinctly separated. The lower chloroform layer containing the lipid was withdrawn and dried over anhydrous sodium sulphate. Lipid solution was filtered quantitatively and the solvent was removed in a rotary vacuum evaporator at 40°C. Lipid thus obtained was transferred with the solvent in a tared conical. Solvents were then removed by a stream of nitrogen gas and finally weighed. Lipid was dissolved in redistilled hexane and stored in a freezer at 18°C. Saponification of total lipid and separation of mixed fatty acids and nonsaponifiables were estimated following Misra ***et al.*** (1984) method. Preparation of mixed methyl esters from free fatty acids was made following Schlenk and Gallerman (1960). Analysis of fatty acid methyl esters was conducted following gas liquid chromatography (GLC). The gas chromatograph used was a Pye Unicam instrument, model GCD, equipped with dual column and dual flame ionisation detector.

Enzymic assay:

For quantitative evaluation of the three major hydrolysing enzymes, five healthy enteropneust worms were selected for the purpose. Four main zones of the gastrointestinal tract viz. (i) branchial, (ii) oesophageal, (iii) intestineI (midgut) and (iv) intestineII (hindgut) were taken into consideration. As the posterior part of the GI tract of living worms was difficult to dissect out, whole body segments were employed for the analyses. All these four zones were cut out from live specimens and immediately preserved in chilled acetone for processing.

For evaluation of enzymatic activity (amylase. protease and lipase) the preserved tissues were freed from acetone in a vacuum desiccator and tissues were homogenised in a temperature below 4°C for one minute in a mechanical homogeniser at 1000 rpm. The enzymes were extracted with double glass distilled water for two hours at4°C.

The homogenate was centrifused at 10,000 rpm for 3 min and the slightly turbid supernatant fluid was used for the assay. Estimation of hydrolysing enzymes in the digestive tract of ***Saccoglossus*** sp. was made following Bernfeld (1955) for amylase. Ichisima (1970) for protease, and Seligman and Nachlas (1963) for esterase. The enzymatic activities were expressed in terms of end product produced per 100 mg of water soluble protein present in the extract. Total protein in the extract was measured following Lowry's method (1951).

Ethological study

The behaviour of the hemichordate worms with the commensals was studied in 'U' shaped glass tubes kept in laboratory aquaria filled with clear interstitial water. At first the worm was introduced in a glass tube along with the polychaete commensals collected from the same burrow. It required one to two days for the animals to acclimatise with the artificially erected habitat and then they exhibited normal behaviour. The intimate proximation between the worm and its commensals could be observed through the transparent glass tube.

SEM study

After proper narcotisation the desired body parts of the worm were cut out and fixed with 2% glutaraldehyde (prepared by 25% glutaraldehyde 1.6 ml phosphate buffer pH 7.2, 18.4 ml) for 2 hours at 37°C in a desiccator. The tissues were then washed through several changes in phosphate buffer solution (pH 7.2) for 45 minutes and then washed with double distilled water for 15 minutes. Then the tissues were transferred to 2% Osmium tetraoxide (OSO_4) in dark for 12 hours till the colour of the tissues turned dark black. The tissues then were washed In double distilled water for 15 minutes and then passed through graded alcohol and after absolute alcohol the tissues were transferred to half acetone and half absolute alcohol mixture for

15 minutes and then the tissues were transferred to pure acetone. After critical point drying (C.P.D.) the tissues were mounted on aluminium stubs and coated with gold (= gold plating) in Edward's sputter coater.

Tissues were finally examined in a Philips Scanning Electron Microscope (P SEM500) at an accelerating voltage of 20 kv. Photographs were taken using 35 mm negative film.

Coelomocyte study

The coelomic fluid together with different types of coelomocytes was drawn out on a slide from the proboscis coelom by cutting the proboscis into two halves. The slide was air dried and stained with Leishman's stain for study under light microscope. Some of the slides containing thin smears were also stained with Giemsa.

Most of the microphotographs were taken in the Marine and Estuarine Biological Research Laboratory. Department of Marine Science, University of Calcutta with the help of Leica M42 Camera and ReichertJung Polyvar Widefield Photomicroscope. Austria. Field Photographs were taken with the help of MX AsahiPentax camera.

Breeding

Mature male and gravid female specimens of ***Saccoglossus*** sp. were collected from the environment and were brought to the laboratory and kept for two or three days in filtered estuarine water. For conducting fertilization in laboratory condition and to study the developmental sequences. Requisite salinity, D.O. and pH of the water were maintained. The developmental study was made during March and April 19831984.

PLATE 3

(A)

Hemichordate bed; Prentice Island, being gradually exposed due to receeding tides.

(B) Author in search of ***Saccoglossus sundarbanensis*** sp. nov. from its habitat.

MORPHOLOGY
MACROANATOMY
EXAMINATION AND DESCRIPTION

Saccoglossus sp. Singh and Choudhury, 1984 a member of the class **Enteropneusta**, phylum Hemichordata and a resident macrobenthos of the mudflats of Sundarbans mangrove ecosystem. displays a long and vermiform body characteristically divisible into three distinct zones, viz. proboscis, collar and trunk. Length of the larger specimens so far unearthed ranges between 35 and 50 cm. Proboscis colour is creamy white, collar is light orange and the trunk is yellowish, in the branchial sector in particular. Sexual dimorphism is distinct in this species (Plates 1 and 4).

PROBOSCIS

This is the most conspicuous first anterior segment of the body and is highly extensible. It measures upto 5.4 cm in the largest living specimens and bears median longitudinal groove or depression throughout its length, both dorsally and ventrally. The proboscis is comparatively tapered at the anterior end. The broader posterior end is connected with the second body segment the collar, by a narrow cylindrical stalk, the neck or proboscis stalk. On the left dorsal side of the stalk is a porethe proboscis pore (Coelomostome). On the ventral surface of the proboscis base leading to the proboscis stalk is a characteristic U shaped depression, the ciliary organ, playing most important role in the filter feeding mechanism of the hemichordate worm. The most important internal anatomical features are accommodated at the rear part of the proboscis and in the proboscis stalk (Fig. 4).

Proboscis coelome

The most diagnostic feature of hemichordate proboscis is its proboscis coelom. The ***Saccoglossus*** sp. under study has paired proboscis coeloms separated distinctly by a thick tissue septum arranged dorsoventrally (Plate 5D & E). In most hemichordates the paired coelomic pouches fuse infront of the stomochord to form a single one at the distal part of the proboscis. At the proximal end of the proboscis the paired coelomic pouches become narrow and enter the proboscis stalk lying dorsal and lateral to the stomochord. The right pouch ends blindly but the left coelomic pauch forms a dilated chamber called the end sac which opens dorsolaterally by a single proboscis pore (Plates 6E & F).

Stomachord

The stomachord is well developed in this species and is formed by the anterior extension of the dorsal wall of the buccal cavity as diverticulum which is evident from the serial transverse and sagital sections. Histologically the stomochord epithelium is very similar to buccal epithelium. Stomochord occupies a position dorsal to the proboscis skeleton. Earlier it was misconceived as 'Notochord' of hemichordates. The stomochord diverticulum extends distally into the proboscis to a considerable length beyond the proboscis skeleton. At its midlength and dorsal to the proboscis skeleton the stomochord becomes very much widened and dorsoventrally compressed (Plates 11 $ 12). Further distally the stomochord gets narrowed gradually and the distal blind end shows a little upturned disposition.

Proboscis skeleton

Proximal part of the proboscis stalk and the roof of the buccal cavity are supported by a chitinous proboscis skeleton. It bears an unusual structural importance in all enteropneusts and is contributed by the basement membrane embedding the nervous layer (Hyman, 1959). The proboscis skeleton consists of a median plate and two posterior extensionshorns or crura (vide Microanatomy page 34) in the shape of an inverted Y (Plate 13A) diverging backwards to be inserted and anchored into the tissue of the dorsolateral wall of the buccal cavity. The two crura form characteristic indentations, usually longitudinally in the buccal epithelium. The median plate bears a median keel and is placed in the proboscis stalk between the buccal epithelium ventrallyand the buccal diverticulum dorsally. From the median plate of the proboscis skeleton series of lateral cartilaginous offshoots are found which after giving secondary branches. curb over dorsally to form complete arches enclosing partially the posterior lobes of proboscis pouches, stomochord and anterior portions of collar coeloms, and sometimes the sinuses. The proboscis skeleton together with the cartilaginous outgrowths probably supports the origin of the extensor longitudinal muscles that are inserted throughout the wall of the proboscis (Plate 14).

In many adult specimens a hard concretion, dark brown in colour. forms a core in the body of the proboscis skeleton. This core is slightly forked behind and extending a little in the respective crura.

Glomerulus

The most conspicuous component of the hemichordate proboscis, apart from stomochord and proboscis skeleton. is the glomerulus which is supposed to be responsible for the removal of the metabolic wastes of the entire animal. It is paired and occupies a ventrolateral position on either side of the stomochord. It is a compact mass of fine tubules (urinary tubules) having simple epithelial lining. This organ is intimately associated with the heart vesicle and is highly vascularised.

COLLAR

Behind the proboscis and partly enveloping it and the proboscis stalk is the collar. a broad and plain bandlike structure without having any wavy margin at its proximal rim. Length of the collar in the largest worm is 4 mm and the diameter is 3.5 mm. At the posterior two third of the collar length there is a concentric ringlike groove.

Oral aperture is situated at the anterior end of the collar on the ventral side of the proboscis stalk and is preceded by the prebral cillary organ on the ventromedial region of the proboscis base. The pair of collar contoms are placed on the dorsolateral side of the buccal cavity separating the dorsal epidermis from the gastric epithelium forming the roof of the buccal cavity. The right and left cavities of the collar coelom are completely separated from each other by the median neural keel, nerve cord, dorsal blood vessel and the paired perihaemal cavities.

Two collar pores (lateral coelomostomes) are present dorsolaterally, one on each side of the posterior margin of the collar. communicating with the respective collar coeloms. Underneath the anterior margins, along the middorsal line of the collar, are present the anterior and posterior neuropores. The ventral mesentery is complete stretching from the mouth to the posterior extremity of the collar.. Dorsally, the dorsal hollow nerve cord is joined throughout Its length with the basement membrane of the epidermis by the neural crest or keel (Plates 16E & F).

TRUNK

This segment of the body is of variable length in different specimens and in very few cases complete animals could be unearthed as the posterior section is very fragile and gets detached because of the mudload gut. It measures in the largest live specimens between 40 to 45 cm from behind the collar to the vent which is terminal in position. The trunk measures 0.45 cm in diameter anteriorly just behind the collar and tapers gradually posteriorly. Anterior two third of the trunk length bears a conspicuous median dorsal groove formed by the right and left elevated and slightly incurved genital ridges or wings (Plate 19D). Middorsal line of the trunk wall is slightly convex bearing the impression of the dorsal epidermal nerve cord. On the proximal part of the body, the ventral side of the trunk is slightly concave, but in the distal sector it is first flattened and then markedly convex. The midventral body wall also shows the impression of ventral epidermal nerve cord.

Proximal part of the trunk is recognised as the branchial sector because of the presence of the series of paired branchial pores on each side. between the genital ridge and the mid dorsal wall of the body (Plates 20C & D). The branchial region measures 3.2 cm long in the largest specimen having 168 pairs of branchial pores (more than any other known species of the genus). In most of the 50 specimens examined so far, the branchial sector measures 2.5 to 2.7 cm in an average and the number of gill pores varies between 85 to 126 pairs. In exceptional cases the number of gill pores exceeds the average count. The gill pores are smaller in the proximal part, increase in size in the middle and again get smaller in size in the distal sector of the branchial region.

Exact extent of the branchial region can only be determined through microanatomical study. Distal part of the trunk is gradually tapered and is dominated by the genital ridges or wings, as is evident from the visibly mature eggs in the gravid females, arranged dorsolaterally in the genital wings ending posteriorly.

Internally, the pharynx is perforated dorsolaterally by paired gillslits which embrace the dorsal two thirds of the pharynx. Ventral part of the pharyngeal cavity is enveloped by the nonbranchial membrane of the pharynx. Two distinct types of gill filaments viz., septal and tongue, are visible in vertical parallel rows. At the posterior extremity of the series the last few gill slits decrease progressively in size but tongues are still present even in the last and smallest one (Plates 13B & 19AE).

On the mid dorsal line of the pharynx there is an epipharyngeal ridge formed of epithelium only. runs continuously upto the oesophageal zone. The ridge seems to be tripartite, the middle one is comparatively larger and hangs freely in the pharyngeal cavity (Plate 20B).

The paired trunk coelomic cavities in the branchial region are almost completely occupied by the gonads, testes or ovaries, thereby totally obliterating the spaces. On the dorsal and ventral sides the almost obliterated trunk coeloms are separated by the dorsal and ventral mesenteries containing blood vessels.

At the Junction of the branchial and oesophageal zone there is a remarkable constriction which demarcates the transition between the two. The worms often get detached at this junction during unearthing. The oesophageal zone is about 2 to 3 mm long. In this zone five to six pairs of intestinal pores open dorsolaterally in the dorsal groove.

Post oesophageal trunk is the longest part of the whole body. Anterior two-third of this body segment bears the evidence of dorsal groove formed by the elevated genital ridges. Beyond this zone the trunk is dorsoventrally flattened or otherwise cylindrical if loaded with ingested mud; the genital ridges are almost absent though rudimentary paired gonads are often visible in histological sections. Posterior part of the trunk is cylindrical and is terminated by the anal opening. The epidermis of this region bears transversely arranged dorsal and dorsolateral papillae and no trace of gonadic elements (Plate 21B).

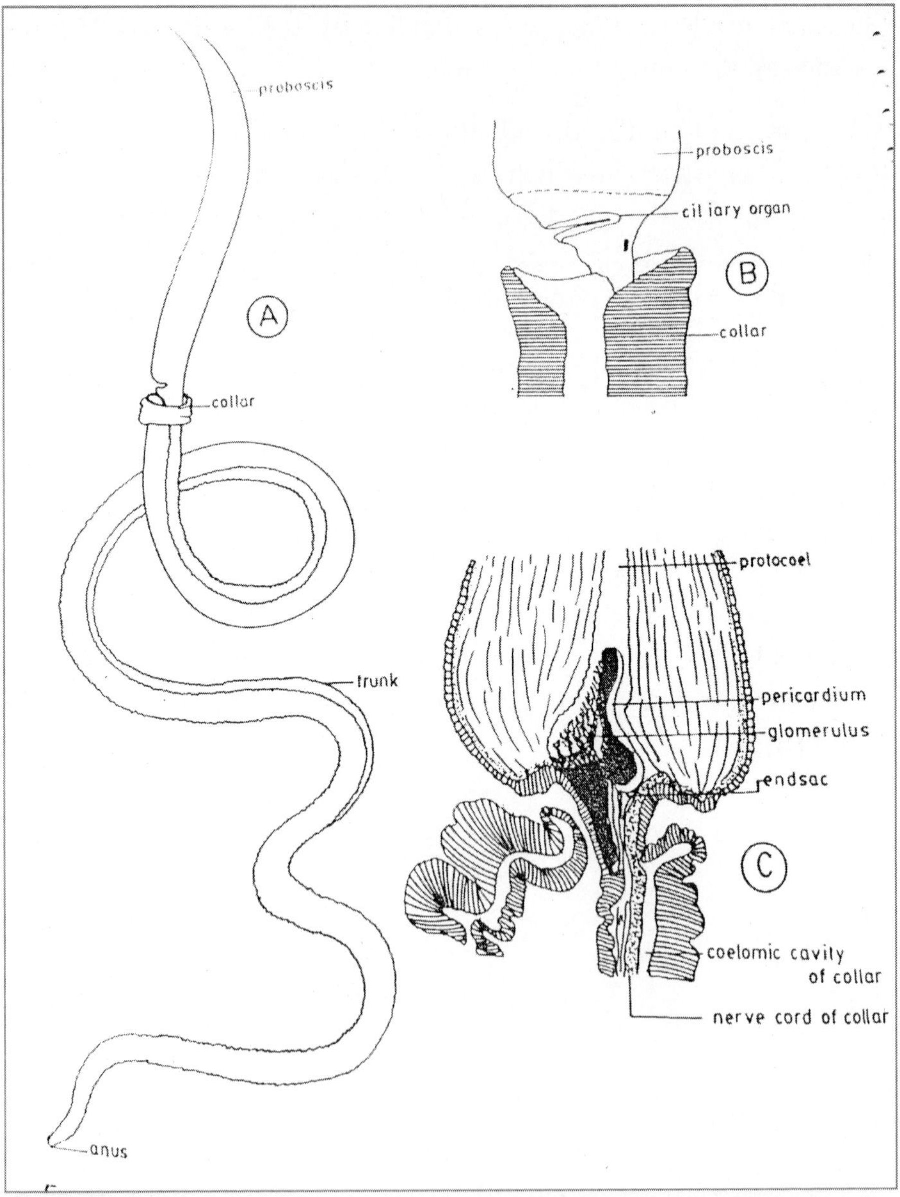

Macroanatomy of Saccoglossus sundarbanesis sp. nov.

MICROANATOMY

PROBOSCIS

Epidermis

Body of the hemichordate worms, enteropneusts in particular, is clothed with a single cell layered ciliated epidermis. This constitutes mostly of tall columnar cells which, ofcourse, do not attain uniform height at different body locations even at different sites of the proboscis itself (Plate 7 AE).

The epidermis of the proboscis in ***Saccoglossus*** sp. is uniformly ciliated throughout and is intersparsed with numerous glandular cells.

In histological preparations numerous nuclei of the crowded epidermal cells appears to be present at all levels except near the surface and near the basal membrane. Surface layer of the epidermis is occupied by rootlet bundles associated with the cilia. The ciliated cells and gland cells have their basal portions constricted to form supporting fibres which remaining attached to the basal membrane, traverse the nerve fibre layer.

The ciliated cells have comparatively large vesicular nuclei which are usually oval, and may occupy different position of the slender cell body. The gland cells, typically of three types, are: mulberry cells, which are sparse, small goblet cells which are abundant, and large mucus cells which are about three times as numerous as goblet cells and half as numerous as the ciliated cells. The size, structure and disposition of the nuclei of the three categories of glandular or secretory cells are also different from each other. Nuclei of the mulberry cells lie at the basal ends, those of the mucus cells being slightly smaller and more deeply situated than those of the ciliated cells. Nuclei of the small goblet cells are still smaller and lie usually near the surface. This observation corroborates the descriptions given for various enteropneusts by Spenge) (1893), Van der Horst (19271939). Brambell and Cole (1936). Bullock (1945) and KinghtJones (1952).

Immediately above the basement membrane and below the ciliated and secretory cells lies the nerve fibre layer. This epidermal nerve fibre layer attains differential thickening at different surface of the proboscis, dorsally, ventrally and laterally, or in the anterior and posterior sections (Plate 10 A & B). This nerve fibre layer is completely devoid of neurons or nerve cells which are present just below the ciliated and secretory cells and above the fibre layer. Nerve fibres that originate from these nerve cell bodies penetrate immediately into the underlying fibre layer (details of Nervous System has been described on page 46).

The nervous layer is well developed in the proboscis of ***Saccoglossus*** species, being thickened to form a cord in the middorsal and midventral groove or depression. The longitudinal fibres of the nervous layer are aggregated into numerous bundles that course along the proboscis toward its tip. This layer also thickens greatly toward the proboscis base, where its fibres tend to arrange in a circular direction and give rise to a definite anterior nerve ring from which the longitudinal bundles take their departure. The nervous layer riches its greater depth on the dorsal and lateral sides of the proboscis stalk and is very thin in the ventral epidermis of the stalk. As in all other ***Saccoglossus*** spp. the dorsal proboscis nerve cord spreads out into a fanshaped area in the posterior surface of the proboscis of the present species.

Ciliary organ

The crescentic groove of the preoral ciliary organ is bounded by a slight continuous ridge formed of a band of cells of greater height with longer and better developed cilia than those of the cells surroundthan those on each side. These tall and conspicuous cells are provided Ing the ciliary organ or lining the groove. That part of the band which borders the outer margin of the ciliary groove tends to be better defined and its cilia are longer than those bounding the inner margin (Fig. 4b and Plate 10 A & B).

The thickness of the epithelial layer of the ciliary organ epidermis varies from 80 to 120 and the nerve fibre layer from 30 to 80 u. This thickness also appears to vary in cross section in the same specimen based on different state of contraction. The epidermis of the preoral ciliary organ is provided with a distinct cuticle approximately 1.5 μ thick and the length of the ciliary tufts borne by the cells varies from 10 p near the margin of the basal region to 35 on the margin of the organ (Plate 10 A & B). The ciliary rootlets are found under high power optical microscope to traverse the thick cuticle.

The number of individual cilia borne by each cell is indicated by the kinetosomes or basal granules of the ciliary rootlets in each cell when examined under high resolution.

Proboscis coelome

The present study and observation reveals that the ***Saccoglossus*** sp. bears double proboscis coeloms that run parallel infront of the stomochord. Longitudinal muscles completely fill up the proboscis accommodating the coeloms which extend considerably toward the anterior tip of the proboscis (Plate 10 C). The coeloms are lined internally by flattened single layered coelomic epithelium (parietal) with an investment of thin circular muscle layer. The paired coeloms are separated from each other infront of the stomochord by a median septum formed of the circular muscles investing the coeloms and being supported by connective tissue which also occurs as packing material in the interspaces of the longitudinal muscle strands of the proboscis, obviously giving passage to blood capillaries (Plate 6 AC).

The proboscis coelom is filled up with coelomic fluid containing floating coelomocytes with vacuolated cytoplasm and relatively small nuclei (Plate 6 D). All other Important organs viz.. glomerulus on either side of the stomochord and the pericardium are lined by thin and flattened coelomic epithelium (visceral).

Muscles of proboscis

In the proboscis of ***Saccoglossus*** sp. the epidermis is immediately followed by a thin layer of circular muscle which is considerably thinner than the nerve fibre layer. Rest part of the proboscis core is packed up by the strands of longitudinal muscles (Plate 10 C) which are supposed to have been originating from the parietal peritoneum lining the proboscis coelom and the median dorsoventral sagital muscle plate present posteriorly near the proboscis complex (Hyman, 1959). This median sagital plate formed of muscle fibres also derived from the parietal peritoneum of the paired coeloms where they meet posteriorly, bifurcates dorsoventrally around the proboscis complex, reunite again ventrally opposed to proboscis ventral mesentery and finally meeting the basement membrane. In cross section muscles are seen to originate from the median sagital plate and to arch above and below in the form of concentric plates possibly supporting various parts of the proboscis complex (Plate 10 DF).

Major strands of longitudinal muscles converge on the dorsal and dorsolateral sides of the proboscis basement membrane at its rear end. On the ventral side the muscle strands converge and insert on the ventral skeletal plate. On the distal part of the proboscis the longitudinal muscle fibres seem to attach to the basement membrane of the epidermis.

Stomochord

From the macroanatomical examinations, at the base of the proboscis, it was revealed that the buccal diverticulum, called stomochord, attained its widest spreading after entering into the proximal part of the proboscis and then gradually narrowed as the diverticulum progresses towards the distal part of the proboscis. At its widest part the stomochord is dorsoventrally flattened and the epithelium forming the floor in this zone is thrown into papillated configurations, both on the right and left flanks of the midline, the roof epithelium being comparatively very thin and smooth. The papillae gradually attained height towards the lateral ends.

The stomochord, as usual is formed of a simple epithelium made up of tall columnar stout cells, but the histological configuration as revealed in longitudinal and cross sections, changes gradually as it progresses distally with the diminution of the stomochord lumen. At one place of the distal end, the epithelial surface membrane bordering the stomochord lumen becomes thick and each epithelial cell surface is found to be projected into denticular peglike processes of undetermined significance (Plates 11 AF and 12 AF).

The epithelial cells of the far distal end are found to be tall and stout coverging towards the lumen and forming a continuous smooth mucous lining. Ultimately the lumen disappears at the stomochord apex.

Heart vesicles and central sinus

Above the stomochord and throughout its length, in the proboscis complex, and below the heart vesicle, lies the central sinus formed of a vesicular structure, rectangular in cross section, containing blood. The lining membrane of the heart vesicle is made up of pavement epithelium possibly contributed by the visceral epithelium of the proboscis coelom. Laterally, the heart vesicle with its contained blood is in direct communication with the digital tubular sinusoids of the glomerulus (Plate 12 AD).

The heart vesicle, also known as pericardial sac, the broad basal part of which rests on the dorsal and lateral sides of the central sinus. Heart vesicle formed of coelomic epithelium, also extends upto the distal part of stomochord. Its roof and lateral walls are made up of pavement epithelium but its broad base resting on the sinus is made up of tall arched cells, in continuation with the coelomic epithelium, and having prominent spherical basal nuclei (Plate 12 A, C, D).

Glomerulus

The bushy and much ramified bilobed glomerulus sheltered within the proboscis coelom, displays the typical histological features of 'renal tubules on either side of the stomochord. The digital tubular processes formed of cuboidal epithelium are supposed to be the contributions of repeated and evaginated foldings of the visceral coelomic epithelium enveloping the stomochord and heart vesicle. The adjoining and juxtaposed epithelial basement membranes of the renal tubules seem to contain inbetween blood spaces or sinusoids, continuous with the central sinus. which are essential for removal of the nitrogenous wastes. At the free distal and lateral ends of the glomerular tubules are found some conspicuous spheroidal cells, some are free, which are supposed to be the chloragogentype of cells, the nephrocytes, responsible for excretion by budding off into the coelom, as is found in the Annelid worms (Plate 9 AF).

Proboscis neck

The narrow proboscis stalk leading to the collar is enveloped by a comparatively thick epidermal epithelium in combination with continuous basal nervous layer. This epithelial blanket encompasses a very delicate and unique neck complex comprised of proboscis coelomic pockets. Perihaemal space and blood sinuses, stomochord and the body of the proboscis skeleton with its mid ventral keel (Plate 14 BF).

From the lateral sides of the proboscis skeleton, less than 1 mm in length approximately, arch out 'cartilaginous' skeletal plates forming semicircular configuration, sometimes completely enveloping and supporting the stomochord, blood sinuses and providing surfaces for insertion of longitudinal muscle bands originating from the proboscis epithelial basement membrane. The conspicuous skeletal arch on each side inturnremify further and reunite on either side to form an irregular lattice work of more smaller pouchlike structures, some twenty in number, housing longitudinal blood sinuses. These sinuses are supposed to have originated from the major perihaemal sinus lying dorsally to the stomochord (Plate 14 B, D).

The proboscis skeleton in the neck region exhibits stratified accretions of chitinous materials which assume a rhomboidal configuration In crosssection. Ventral to the body of the skeleton is the keel which is also made up of concretions of chitin (Plate 14 B. DF). The stomochord or gut diverticulum. lying just above the proboscis skeleton. and throughout its length, is uniformly semicircular in cross section. The lumen of the stomochord is guarded by the tall columnar epithelium with differential depth in the roof and floor and on the lateral edges. Conspicuous spherical nuclei always border the apical ends of the columnar cells which are devoid of cilia in this region (Plate 12 A).

Coelomic pouch in the proboscis stalk is represented by the left dorsal proboscis coelomic pocket and a pair of ventrolateral coelomic pockets formed of tall columnar epithelium having terminal nuclei. Dorsal coelomic pouch is situated partly in the dorsolateral position. on the left side of the proboscis stalk and dorsal to the perihaemal space. It is actually the representation of left dorsal coelomic pouch. Ventrolateral coelomic pockets are circular in crosssection and are located just below the proboscis skeleton. These are the blind ends of the paired proboscis coeloms which extend far behind the proboscis neck (Plate 14 C).

The left coelomic pouch on the dorsolateral side that forms the dilated chamber called end sac, is lined by simple ciliated epithelium formed of columnar cells of variable depth displaying conspicuous apical nuclei. This epithelium is in continuation with the proboscis neck epidermal epithelium invaginated at the confluence of the proboscis pore. Thick sheet of subepidermal nervous layer is present on the dorsal and lateral roofs of the end sac. Right basal part of the end sac is occupied by the perihaemal space of the collar impregnated with longitudinal muscles of the proximal part of the trunk. The left basal part of the end sac is occupied with the thick band of longitudinal muscles of the proboscis, supposedly forming the sphincter of the tubular proboscis canal leading to the proboscis pore (Plate 6 E, F).

COLLAR

The collar epidermis is differentiated microanatomically in six distinct histological zones each containing ciliated elements and glandular cells in varying proportion. The first zone comprising the anterior margin of the collar is formed mainly of tall columnar ciliated cells and does not possess any glandular cells. The epithelium of this zone is comparatively thinner than the other collar zones. At the middorsal region of the proximal collar depression where the proboscis epithelium merges with the 1st zone is the opening of the anterior neuropore.

The second zone is the thickest zone and mostly occupied by deeply staining glandular cells. This zone extends about two third of the dorsal collar length and the chromophilic substances are apical in position. This zone is sharply demarkated by its chromophility from both the first and third zone and increases in thickness posteriorly. The second and third zones are separated by a deep vertical notch or indentation. The third zone is not so densely chromophilic but harbours sufficient number of mucous secreting glandular components at the distal ends. This zone gradually merges with the fourth which is equally thick like the former having tall columnar cells. This zone seems to be completely devoid of any glandular elements.

Between the fourth and fifth zone there is a small indentation which marks the transition to fifth zone. This collar zone is comparatively smaller in thickness and like the fourth zone is almost completely devoid of glandular elements, but highly ciliated. The fifth zone is followed by the sixth zone which bends vertically, at right angle to the body axis, and marks the last zone of the collar in this enteropneust species. The epithelium of the sixth zone is thicker at its anterior end and gradually narrows at the collar trunk transition.

Mucus secreting cells are sparsely distributed at the distal part of the entire epithelium.

Paired collar coeloms rest on the dorsolateral side of the widened nerve cord and are packed up with coelomocytes. On the dorsal periphery of the parietal epithelium of collar coeloms are found, in cross sections, series of peripheral longitudinal blood vessels just beneath he thin sheet of nervous tissue forming the basement layer of the collar epidermis (Plate 16 A. B). Collar coeloms of both the sides extend ventrolaterally on either side of the buccal cavity and meet midventrally to form the ventral mesentery through which passes the ventral vessel. Major part of the coelomic cavities are filled up with longitudinal and circular muscles contributed both by the collar and trunk. These muscles take different courses horizontal. vertical or oblique to end at their point of Insertion and hence pose delicate and confusing microanatomical configurations when cross sections of the regions are examined.

Dorsal nerve cord in the collar is very thick and subepidermal in position. In cross section nerve cord is found to be dorsoventrally compressed and forms a very conspicuous anatomical feature in the enteropneust hemichordates. A distinct neural canal is present in the anterior half of the collar cord but is absolutely solid in the posterior half till it enters the trunk as epidermal neural cord. Ventral half of the nerve cord is formed of nervous tissue fibres and the dorsal half is occupied with many layered neurons (Plate 15 E).

Microscopical studies of the anterior and posterior part of the collar reveals the presence of anterior and posterior neural pits underneath the collar epidermis but these are not in direct communication with the neural canal. In the anterior sector, dorsal and dorsolateral collar flap has been invaginated so much so that a deep crescentic notch is formed between the collar and the proboscis neck to form the neural ptt. In cross section of this part a complete section of this neural pit of epidermal origin is found to be entirely Incorporated and housed between the collar epidermis, collar coeloms and the dorsal nerve cord (Plate 16 C, D). From the mid dorsal region of the nerve cord, throughout the length of the collar, rises the neural keel separating the two collar coeloms and to contact the epidermis dorsally. The neural keel is formed by the Juxtaposition of the peritoneal membranes of the right and left collar coeloms Incorporating neural material. The keel peritoneal membranes display a peculiar pavement epithellum with conspicuous nuclei (Plate 16 E, F).

Below the dorsal nerve cord is found the paired perihaemal spaces, also laterally extended and resting on the dorsal part of the buccal epithelium. The membrane lining the perihaomal spaces is also formed of coelomic epithelium (but not of collar coeloms) contributed by the collartrunk septum. Inside the perihaemal space transverse section of longitudinal muscle fibres are found. The two perihaemal spaces associate medially to form dorsal mesentery accommodating the dorsal subneural blood vessel (Plate 15 AF). Below the perthaemal spaces lies the buccal epithelium which is formed of tall columnar cillated cells having apical distinct nuclei. Columnar epithelium of the floor of the buccal cavity shows irregular foldings and is much more thicker than the roof (Plate 18 A).

Between the peritoneal lining of the buccal epithelium and the collar epidermis on the ventrolateral sides are found two layers of longitudinal muscle bands enclosed within and limited by the perito real membranes of two sets of obliquely disposed ventrolateral coelomic spaces, collar proximally and trunk distally, contributed by the collar trunk septum. The longitudinal muscles contributed by the trunk are extended anteriorly as peribuccal muscles, turn upward to be inserted on the posterior surface of the coelomic epithelium (of the trunk) that is placed almost in the vicinity of the cornue of the proboscis skeleton. Longitudinal muscles of the collar are inserted on the proximal surface of the collar coelomic epithelium. At the anterior margin of the oral aperture a thin sheet of circular muscles is present in the collar coeloms associated with the visceral peritoneum supporting the buccal epithelium.

TRUNK

Branchial sector

In transverse section the trunk epidermis in the branchial sector, is comparatively thick in the ventral and ventrolateral margin than the dorsal side. The epidermis in glandular and cillated throughout its perierter but more glandular at the dorsolateral side where the branchtal pouches open through the branchial pores. Just below the epidermis is the band of longitudinal muscles showing differential thickness at different region of the trunk. The muscle bands display substantial thickening at the ventral and ventrolateral sides supporting the branchial pouches (Plate 20 AC).

The epidermis in the present hemichordate species is slightly elevated in the mid dorsal line rendering it a noticeable convex configuration which accommodates the dorsal solid nerve cord which is physically embedded in the integument forming its basal part. On the ventral side of the trunk, the ventral nerve cord is also epidermal and physically associated with it like the dorsal nerve cord. Ventral nerve cord is triangular in cross section, predominantly formed of nerve fibres and very little neurons. Subepidermal thin nerve layer is extended to some extent on the right and left flank of the ventral epidermis.

Underneath the epidermis and on either side of the ventral blood sinus well developed longitudinal muscles are present ventrally and laterally around the pharyngeal apparatus and the gonads. Longitudinal muscles are also present on the dorsolateral side under the dorsal epidermis where it lines the dorsal groove between the two genital wings or ridges. Very thin patches of circular muscles are found to be present between the longitudinal muscle and the ventral outer wall of the pharyngeal endodermis. Circular muscle is absent from the rest part of trunk in the branchial sector.

Trunk coelom is restricted in the genital wing where its existence can be perceived as the limiting membrane accommodating the gonadsitestes or ovary). On the middorsal and mid-ventral position of the trunk the almost obliterated coeloms are represented as the peritoneal membranes associating to form the dorsal and ventral mesenteries sheltering the dorsal blood sinus and ventral blood sinus respectively. Gonads are saccular or lobulated being separated by thin sheet of connective tissue membranes. In different individual sections. gonads of either sex are found to display sperms or eggs at different stages of their maturity or development (Plate no 22 &23).

Branchial cavity to this trunk sector is lined by endodermis formed of tall columnar ciliated epithelium with distinct apical nuclei. Endodermis of the middorsal roof of the pharyngeal cavity is modified to a leaflike configuration in cross section called the dorsal "ridge' or epibranchial ridge which seems to be hanging throughout the length of the pharyngeal roof. The dorsal ridge is formed of comparatively taller columnar cells than the rest part of the pharyngeal epithelium (endodermis).

In the serial sections of the branchial sector at places where the pharyngeal wall is perforated by the branchial slits, the branchial lamellao (primary and secondary or tongue septum) are found to be supported internally by stiffened noncellular skeletal mass, seemingly chitinous in character, called branchial bars. The epithelium lining and wrapping the branchial lamellae is conspicuously modified from the adjacent pharyngeal epithelium with which it merges all the time. Lamellar epithelia on the lateral sides are strongly ciliated and are densely packed up with deep staining nuclei at different strata giving it a stratified character (Plate 20 C)..

Oesophageal sector

Trunk epidermis in this sector is very characteristic displaying tripple folds in the dorsal groove between the genital ridges. Middorsal fold accommodates the neural cord and the epithelial cells are mostly vacuolated and glandular (mucold). Epithelium of the two other adjacent folds, right and left of the median fold, are also glandular.

possibly mucold, and is in continuation with general trunk epithelium that envelopes the genital ridges and the ventral part of this trunk segment. On the midventral region where the two coelomic peritoneal associate to form the ventral mesentery, the ventral nerve cord, triangular in cross section, is present just below it and in intimate association with the ventral epidermis.

Subepidermal longitudinal muscles form two thick distinct lobes on either side of the ventral mesentery, then gradually lose their thickness, as they ascent ventrolaterally to envelop the dorsolateral gonadal ridges. Very thin sheet of longitudinal muscle is present below the dorsal epidermis that provide noticeable support to the dorsal mesentery and partly envelop the dorsolateral part of oesophageal roof. Circular muscle is almost absent in this trunk sector. Gonadal ridges are very prominent here that display distinct lobes (6 to 9) on each side being compartmented by the invasion of the coelomic peritonial membrane associated with longitudinal muscles and blood. sinuses (Plate 23 A).

Castric epithelium (endodermis) of the oesophagus is represented by tall columnar ciliated cells which at places become very thick partly obstructing the gut lumen so as to play the role of a sphincter as it is supported by both circular and longitudinal muscles. At the transition of the pharynx and oesophagus the last rudimentary pharyngeal slits, deeply compressed. render a very astounding histological configuration on dorsal side of the oesophageal lumen where the oesophagus is divided into regions with intestinecutane furrows, possibly leading to intestine (Plate 20 E).

Mid trunk sector

In cross section microanatomy of this zone of the trunk is almost same to that of the branchial zone so far the epidermis, nerve cords, musculature and gonads are concerned. The gut endodermis is the only exceptional structure which retains its regional gastric identity. There is no epibranchial ridge in this sector, instead uniform

Involutions of the gastric epithelium are met with here, ofcourse with certain conspicuousness on the left ventrolateral side of the gastric lumen. Cells of this zone also deserve some special demarcation from the rest part of the lumen which are attributed to share the responsibility of the hepatic function.

Hind trunk sector

Hind part of the trunk is narrow and tubular and without dorsal and ventral grooves. Gonadal ridges are almost reduced showing rudimentary gonads in sections. Epidermis is predominated with mucoid cells. supported underneath on the dorsolateral and ventrolateral sides by the strong lobes of longitudinal muscles which are considerably reduced on the lateral sides. Circular muscles are completely absent in this trunk zone. The most conspicuous feature of this sector is the gut endodermis which is enormously thick with much widened ciliated columnar epithelium. Tremendous foldings of the gut epithelium reduces the lumen of the trunk sector which is responsible for the formation of faecal pellets. At the anal part, the closely set dorsal and dorsolateral epidermal papillae are found to be formed of tall epidermal glandular cells.

Gonads

Ovaries and testes are lodged in the lateral trunk coeloms, dorsolateral in disposition and beginning from the very first trunk segment of the branchial zone. just from behind the collar, continue throughout the length of the trunk and cease to exist a little ahead of the anal end.

Ovaries are supported by epidermis and longitudinal muscles of the dorsolateral genital ridges. They are enveloped by coelomic epithelium which at places ingresses transversely or obliquely into the gonadic material to divide it in different lobes of different dimensions. These invaginated coelomic epithelia carry with it part of longitudinal muscles, connective tissue and blood sinuses. From the micro anatomy of the gonadal ridge, it appears that the inner surface of the coelomic epithelium (probably visceral) germ mother cells are proliferated which ultimately give rise to the primary oocytes. Size of the primary oocytes ranges between 30 to 200 pm. Mature oocytesare released as potential eggs which experience further meiotic divisions during the process of fertilization outside the parent body.

Female gonopore: There is no permanent gonoduct and hence no permanent gonopore in this species but during the breeding season definite exit passage develops to release eggs with very specialised histological characteristics leading to an opening in the dorsal groove on either side of the middorsal line. In females, at regular intervals in the trunk segments, the temporary exit pores, known as gonopores. are guarded by very specialised vacuolated cells, probably of coelomic origin, arranged and aggregated in the form of a buldged dome acting as 'plugs' (Plate 22 A & B). Just below this plug and on either side of it. two narrow liplike structures are present formed of longitudinal muscles and coelomic epithelium.

Testes in male specimens occupy similar positions in the genital ridges and exhibit similar supporting tissue elements like the female gonads. Each testis is secondarily sacculated or lobed by the ingression of the epithelial partitions evolved from the coelomic wall. Each male gonad or testis narrows to a neck that opens externally by the male gonopore in the branchiogenital groove. Spermatocytes at different stages of development including mature and finished sperms have been observed in the histological sections of the male gonads (Plate 23. AE).

In histological sections of both male and female gonads, large opaque bodies are visible at immature stages, particularly in the pre breeding season. Sections of the gonads during breeding and post breeding seasons do not show the presence of these globules, thereby indicating that these opaque bodies might be some storage nutrients which are used up during the process of maturation.

Salient features and biological system of ***Saccoglossus sundarbanensis*** sp. nov.

External features

The cylindrical rather flacid worm like body of ***Saccoglossus sundarbanensis*** sp. nov. displays typical enteropneust tripartite body configuration composed of an anterior proboscis, an intermediate collar, and a long trunk. These features correspond to the salient deuterostome body divisions protosome, mesosome and metasome.

The proboscis or protosome is exceptionally elongated in the present ***Saccoglossus*** species with conical and pointed distal tip. It is generally dorsoventrally flattened bearing a deep middorsal groove and a midventral depression at the proboscis base. At its rear end. the proboscis narrows to the slender proboscis stalk.

The collar or mesosome is a short cylindrical body segment. the surface of which is often marked with elevations, depressions, and specially circular grooves after fixation. The funnel-like anterior part of the collar is known as collarette which embraces the proboscis stalk and also the posterior part of the proboscis.

The conspicuously long trunk or metasome is dorsoventrally flattened and bears a middorsal and a mid ventral ridge for accommodation of the corresponding median longitudinal nerves and blood vessels. Major part of the trunk length from behind the collar is laterally frilled representing pseudo segmentation. Proximal part of the trunk in this species, also from behind the collar, displays paired genital wings thus giving effect to a dorsomedian groove like configuration in which open the ganopores and the gillpores. The genital wings gradually diminish distally and the trunk attains more or less a rounded feature till it terminates to the anal end.

Body wall and integumentary system

Tripartite soft body of the enteropneust worm is clothed with ciliated epidermis formed of tall columnar epithelial cells. The height of the cells, ofcourse, are variable regionally and are interspersed with gland cells, richly so in most part of the body.

Unlike other coelomate invertebrates, the integumental epidermis of the ***Saccoglossus*** species is conspicuous and complicated in the sense the epidermis passes indistinguishably into a thick nervous layer that occupies its base and is traversed by the filamentous bases of the epidermal cells.

Nervous system

Nervous system examined and studied in ***Saccoglossus sundarbanensis*** by the present author corroborates with the excellent consideration of the enteropneust nervous system as presented by Bullock (1945). Silen (1950) and KnightJones (1952). Like other enteropneust members the nervous system is in a very primitive level of its organization. In different regions of the body, the nerve plexus at the base of the surface epithelium has become thickened to form longitudinal epidermal nerve cords. The principal nerve cords are the midventral and middorsal of the collar and trunk. The ventral trunk cord terminates at anterior end of the trunk. The dorsal trunk cord continues into the collar but at once leaves the epidermis and runs forward through the collar coeloms above the buccal tube and above the perihaemal canals. This actually becomes internal in position and is called the collar cord (neurocord of KnightJones) when present as a pronounced longitudinal thickening.

In this species, as in some other enteropneust worms, the collar cord is hollow and is perhaps homologous to the dorsal hollow nerve cord of chordates. The collar cord is continuous with the general epidermal nerve plexus at both ends. As this worm lacks the usual characteristics of a brain. the collar cord presumably represents the nervous center, resembling a conduction path. The collar cord possesses giant nerve cells apparently conducting impulses from proboscis to the collar and trunk.

In the proboscis, one particular prominent strand forms the so called dorsal nerve cord but many more smaller nerve strands may be seen in transverse section. All of these run forwards from the anterior nerve ring with a posterior thickening at the proboscis base, formed mainly of nerve fibres running on a circular course around the proboscis.

The dorsal and ventral nerve cords in the trunk run forward to join the peribranchial (= circumenteric) nerve ring and form the chief nervous link between those two regions. The peribranchial nerve ring is a thickening formed of bundles of fibers that run in a circular course across the boundary between the collar and the trunk (Kinght & Jones, 1952).

Digestive system

In conformity with the general enteropneust pattern, the digestive tract in the present hemichordate worm is an epithelial tube which runs straight from mouth to anus and in general lacks intrinsic musculature. The tract begins with a large mouth that occurs ventrally in the collar between the collarette and the proboscis stalk. The mouth leads into the buccal cavity that occupies the interior of the whole collar length and is lined with a very tall, vacuolated, ciliated epithelium, often intersparsed with goblet type gland cells and having a nervous layer in its base.

From the roof of the buccal cavity evolves the much disputed which begins as a narrow neck and then passes anteriorly just above But diverticulum (buccal diverticulum or stomachord or 'notochord') the main plate of the proboscis skeleton to reach the posterior part of the proboscis. There it participates in the 'proboscis complex' projecting into the proboscis coelom. Epithelial wall of the diverticulum resembles buccal cavity in having tall, vacuolated epithelial cells in which cilia and gland cells are often visible.

Following the terminology used by Van der Horst (1939), the first part of the digestive tract in the present species is also designated as the pharynx as this part bears the paired gill slits or apertures which, of course, do not occupy the entire circumference of the pharynx. Usually the pharynx is regionated into a dorsal respiratory half perfora ted by gill slits and a ventral half that serves digestive function. The epithelium of the later may not differ from that of the buccal tube and is ciliated and provided with numerous small gland cells. The branchial part of the pharynx gradually diminishes and passes smoothly into the oesophagus after the last gill slits showing distinct regionation. A remarkable feature of the oesophagus in this species is the presence of oesophageal canals and pores by which the oesophageal lumen communicates dorsally with the exterior. These usually occur in pairs in the posterior thinner part of the oesophagus.

Beyond oesophagus. the remaining part of the digestive tract in the trunk is termed intestine. In ***S. sundarbanensis***like all other members of the genus ***Saccoglossus*** there is no external manifestation of hepatic region in the form of dorsolateral sacculations. Histological investigation shows that the anterior darker region of the right side of the intestinal epithelium contains numerous tall gland cells, wanting in the posterior paler region or in the left side of the anterior intestinal epithelium. This glandular region may serve the purpose of hepatic functions in this particular hemichordate specimen.

Beyond this so called hepatic region the intestine continues as a tube formed of tall ciliated epithelium with a moderately narrow lumen, terminating ultimately in the anus.

Respiratory system

Pharynx, the prime apparatus for respiration in enteropneust worms, occupies a considerable length of the postcollar trunk segment. The most significant diagnostic feature of the hemichordates is the development of pharyngotremy i.e., the branchial region perforated on each side by a longitudinal series of openings, the gill slits, oriented dorsoventrally and leaving a narrow middorsal pharyngeal strip. (epibranchial ridge) between their dorsal ends. The gill slits do not open directly to the exterior, but instead into pouch like cavities, the branchial sacs, that in turn open on the dorsal body surface by the gill pores.

The number of gill pores ranges in ***S. sundarbanensis***from 85 to 168 pairs, the number increases possibly with the age and length of the individual animal. According to Barrington (1965), pores are formed throughout life at the posterior end of the series, thus lengthening the branchial region. The gill pores are provided with sphincter muscles and typically the row of gill pores on each side is sunk into a groove, the branchiogenital groove, since it also received gonopores.

The pharyngeal gill slits in the anterior trunk region of the worm are responsible for gas exchange. Each slit opens through the side of the pharyngeal wall as a "U" shaped cleft. The pharyngeal wall between the clefts and that part of the wall projecting downward between the arms of the "U" (the tongue bar) are supported by skeletal thickenings (bars) of the basement membrane of the pharyngeal epithelium. The septa and tongue bars are ciliated on both the pharyngeal and lateral faces and internally contain a plexus of blood sinuses from the ventral longitudinal vessel. The beating of the cilia of the tongue and septal bars drag the stream of water from the mouth and passes out through the gill slits.

Blood vascular system

The haemal system of the hemichordate worm under investigation displays, as in all other enteropneusts, a complex set of lacunar or haemocoelic spaces compounded with definite vessels and having a propulsatory center in the proboscis complex. There are two major vessels. the dorsal and the ventral longitudinal vessels, the former enlarges Into a venous sinus anteriorly and is partly surrounded by the wall formed by the paired perihaemal cavities also known as heart vesicle which contain muscles.

The dorsal vessel runs in the dorsal mesentery below the dorsal nerve cord from the anus to the collar, where it descends slightly to occupy a position between the two perihaemal cavities (Plate 15, A-E)

The venous sinus receives from in front a pair of lateral proboscis veins, one from each side of the proboscis (Plate 7, D). The venous sinus then passes anteriorly in the proboscis base, directly above the stomochord and below the heart vesicle to form an elongated expansion the **central sinus**, often called heart. The heart vesicle also called pericardial or cardiopericardial sac (Hyman, 1959), is a sac rectangular in cross section whose widened base rests directly on the central sinus (Plate 12, A, C & D). From the central sinus blood is transmitted directly into the glomerulus on either side of the proboscis complex.

The blood leaves the glomerulus by way of four vessels, two of them supplying the proboscis as a middorsal and midventral proboscis artery. The latter is larger which descends along the anterior edge of the ventral mesentery of the proboscis where it profusely ramifies along the ventral proboscis wall. The smaller mid dorsal artery ascends and branches in the dorsal proboscis wall. **Two other efferent glomerular** arteries carry blood from the glomerulus, one on each side of the buccal diverticulum (stomachord). turn ventrally to encircle the buccal tube as the peribuccal vessels only to meet ventrally below the buccal tube in the collar to form the ventral longitudinal vessel. A ring vessel originates from the ventral longitudinal vessel that traverses the collar trunk system and supply the collar tissues through two extensive lacunar networks. The dorsal vessel gives rise to a rich lacunar network around the dorsal collar nerve cord that may or may not concentrate dorsal to the cord to an epineural vessel.

Ventral longitudinal vessel passes posterial imbedded in the ventral trunk mesentery to reach the anal end. During its course the vessel Is in intimate contact with the ventral body wall on one hand and the digestive tube on the other, and supplies both of them through lacunar networks lying between the two lamellae of the basement membranes. In the branchial segment of the trunk the branchial apparatus Is supplied with the vascular plexus of the pharynx contributed by the ventral vessel. The complex architecture of the branchial vascularisation is such that each septal and tongue bars are supplied with afferent or arterial vessels which break up into plexuses. Efferent or venous vessels reform from the branchial plexuses and ascent dorsally to meet with the dorsal longitudinal vessel. The gonads are supplied from the epidermal vascular plexus throughout the length of the trunk. It has been noted that the peripheral blood vascular system in this ***Saccoglossus*** sp. Like other enteropneusts, are comprised of mere lacunar spaces instead of copillaries. The circulatory system in hemichordates, therefore, is primarily of open type. The blood is colourless and seems to be without any blood corpuscles. Blood flows forward In the dorsal longitudinal vessel and backward in the ventral longitudinal vessel.

Excretory system

The glomerulus, supposed to be the only essential organ for excretion in acorn worms, constitutes most conspicuous and significant component of the proboscis complex in ***S.sundarbanensis***. This complex is accommodated in the posterior part of the proboscis coelom, ventrolaterally to the proboscis skeleton. In ***S. sundarbanensis***the glomerulus is represented as a paired organ, displayed on either side of the stomochord as profusely evaginated digital processes from the central sinus, lined by the peritoneum of the proboscis coelom. Glomerular digital processes are actually fine blood sinuses and all blood from the central sinus is driven anteriorly through the glomerular sinus system. Very little is known about the excretory role of glomerulus in hemichordates.

Reproductive system

All enteropneusts are dioecious, so is the present ***Saccoglossus*** species. Sexual dimorphism is not very conspicuous but becomes evident during the breeding season when gravid females can be sharply distinguished from the males by changed colour and characteristics of gonadal ridges. Gonads, ovaries or testes, are found to be present in this species right from the very first trunk segment just behind the collar and continued almost to the end of the trunk, a little distance from the terminal anus.

There is no permanent gonaduct but during the breeding season definite exit passages develop in respective gonads to release eggs and sperms to the exterior. The passages, known as gonopores bearing very specialised tissue characteristics, open beside the branchial or gill pores in the branchiogenital grooves on either side of the mid dorsal convex ridge of the trunk.

DISCUSSION

The author enjoys the privilege to communicate ***Saccoglossus*** sundarbanensis sp. nov. in the present treatise as a new and the fourteenth representative member in succession of the genus ***Saccoglossus*** under the phylum Hemichordata (Class: **Enteropneusta**, Family: Harrimaniidae). The author also enjoys the priority to report a hemichordate from any mangrove environment of the tropical world.

The family Harrimaniidae was erected by Spengel in 1893 sheltering those enteropneusts lacking lateral hepatic sacculations, genital wings. branchial synapticulae and lateral septa. The genus ***Saccoglossus*** was first erected by Schimkewitsch in 1892 but remained unnoticed till 1934 when van der Horst revived the name of the genus to accommodate those enteropneusts having (1) very long proboscis, (ii) collar about as long as broad, (iii) lateral genital ridges and (iv) many small gill pores but distinct.

The hemichordate worm protrayed in the present dissertation satisfies all the characteristic features necessary for its inclusion in the genus ***Saccoglossus*** and the family Harrimaniidae. But the worm differs in several points from the published descriptions of the thirteen other species of the genus and displays very exclusive diagnostic features of its own. Hence the author intends to propose and erect a new species (fourteenth in order) under the genus ***Saccoglossus*** specially to shelter the worm under discussion, Before finalising Its specific identity let it be distinguished from the other species on the following combination of characters:

(1) ***Saccoglossus kowalevskyi*** (Agassiz) locality: Atlantic coast of U.S.A. A hundred pairs of gill pores: genital pores begin one millimeter behind the collar: only four or five concentric ringsare clearly visible in the longitudinal muscles of the proboscis; four to six

pairs of intestinal pores; small proboscis coelom; and presence of an 'operculum formed by the posterior margin of the collar covering the first gill pores.

(2)***Saccoglossus mereschkowskii*** (Nic. Wagner). Locality: North 1895 Eastern Russia. 50 pairs of gill pores; arrangement of the longitudinal muscle fibres of the proboscis in seven to ten concentric rings: absence of ventral mesentery of the collar anteriorly: tongue lamellae do not project beyond the septa in lumen of the pharynx: end plate of the proboscis skeleton bears a long dorsomedian spine; seven pairs of intestinal pores; presence of definite genital ridges.

(3)***Saccoglossusinhacensis*** (Kapelus. 1936). Locality: South East Africa. Eighty two or more pairs of gill pores; absence of tongues in the last pairs of gill slits; longitudinal muscles in the proboscis not in concentric rings; gonads begin at the level of the fourth gill pores; four pairs of intestinal pores: the first of which has four internal openings: absence of dorsal mesentery in the collar; direct continuity between the neural crest and epidermal nerve layer of the collar without the intervention of a basement membrane.

(4) ***Saccoglossus pusillus*** (Ritter). Locality: California, U.S.A. About sixty pairs of gill pores and one pair of intestinal pores; crura or horns of the proboscis skeleton extend to the hinder end of the collar and embrace threequarters of the circumference of the buccal cavity.

(5)***Saccoglossuscaraibicus*** Van der Horst, 1924. Locality : Indies. West Gill pores more than fifty pairs; longitudinal muscles of the proboscis not in concentric rings: proboscis pore median in position: great development of the circular musculature ofthe collar which is twice as thick as the epidermal nerve layer: ventral proboscis septum extended to the tip of the stomochord: glomerulus dorsally placed; gonads begin between fourth and fifth gill pores.

(6)***Saccoglossusbournel*** (Menon, 1904). Locality: Madras. India. Longitudinal muscle fibres of the proboscis are not arranged in concentric rings: dorsal mesentery of collar complete; ventral collar septum absent anteriorly; crura or posterior horns of the proboscis skeleton extend to the hinder end of the collar; more than 62 pairs of gill pores; gonads beginning immediately behind the collar: very deep anterior neuropore.

(7) ***Saccoglossus sulcatus*** (Spengel, 1893). Locality : Japan. Deep dorsal sulcus on the proboscis rendering it crescentic in cross section: ten to eleven pairs of gill pores; other characters not known.

(8)***Saccoglossusruber*** (Tattersall, 1905). Locality: Western Ireland. Longitudinal muscles of the proboscis not in concentric rings: no genital or muscular ridges; fifty six to sixty four pairs of gill pores (sixty to ninety pairs of gill pores in S. cambrensis which has been synonymised with S. ruber); wide proboscis coelom; S-shaped bend of stomochord infront of ventral diverticulum.

(9)***Saccoglossusgurnel*** (Robinson, 1927). Locality: Suez. Collar nearly twice as broad as long; longitudinal muscles in concentric rings; forty to sixty pairs of gill pores: ventral septum of proboscis very long, extending beyond the tip of stomachord: dorsal diverticulum of buccal cavity present: median proboscis pore: gonads beginning immediately behind collar; dorsal and ventral mesenteries of the collar complete throughout 4/5 collar length.

(10)***Saccoglossusatagensis*** (Benham, 1930). Locality: New Zealand. Deep dorsal groove on the proboscis, longitudinal muscles of the proboscis in three or four concentric rings: ten to fifteen pairs of gill pores; one pair of intestinal pore: gonads extend anteriorly to the level of fourth gill pore: dorsal and ventral mesenteries of collar absent; dorsal diverticulum of buccal cavity present.

(11)***Saccoglossuspygmaeus*** Hinrichs and Jacobi, 1938. Locality : Heligoland. Longitudinal muscles of proboscis not in concentric rings: proboscis without pronounced dorsal groove; nine to twenty two pairs of gill pores; dorsal diverticulum of buccal cavity absent: dorsal and ventral mesenteries of collar complete; gonads begin at posterior extremity of the branchial region of trunk: ventral septum of proboscis very short; one pair of intestinal pores: very small in length, about 3 cm.

(12)***Saccoglossusapantesis*** Thomas, 1956. Locality: South Australia. Longitudinal muscles of the proboscis is arranged in at least nine or ten concentric rings; thirty to forty five pairs of gill pores: gonads start in the mid branchial region of the trunk and form marked dorsomedial ridges; ventral longitudinal muscles of the trunk form distinct ventrolateral ridges into the posterior branchial and oesophageal regions; six to eight pairs of intestinal pores: collar epidermis has five distinct zones. the second of these being almost equal to the remaining four added together: the stomochord has a ventral diverticulum which is directed slightly backwards and is partially grasped by the bluntly bifid tip of the proboscis skeleton: crura of the proboscis skeleton extend almost half way of the collar length and embrace slightly more than half the circumference of the buccal cavity.

(13)***Saccoglossushorsti Brambell and Goodhart***, 1941. Locality : Southern England. Dorsal and ventral grooves on the proboscisextend throughout its length: collar forms a slight operculum covering a part of branchial trunk: longitudinal muscle fibres of the proboscis arranged in nine or more concentric rings; hundred to hundred forty pairs of gill pores; gonads beginning within 1 mm of the collar; rounded genital ridges present: stomochord straight with ventral diverticulum and wide lumen through out: no dorsal glomerulus; single proboscis pore on the left side: proboscis skeleton embracing half to two thirds of the circumference of the buccal cavity and extending one fourth to one third of the length of the collar: collar epidermis has four distinct zones: four to eight intestinal pores; gonads not lobed.

Apart from these thirteen species. two more species were described and reported under this genus. These were ***S. cambrensis*** Brambell and Cole (1939) and S. serpentinus (Assheton, 1908). In their revision work on the genus ***Saccoglossus*** , BurdonJones and Patil (1960) compared at length these two species with S. ruber (Tattersall. 1905) and ultimately synonymised them with the latter thus reducing the total number of species under the genus from fifteen to thirteen. They also ommitted to include in their monograph, perhaps by unintentional mistake, another valid species, ***S. apantesis*** reported from Australia by 1.M. Thomas in 1956 and so the total number of species under the genus remained confined to twelve instead of thirteen.

The first hemichordate worm reported from India under the genus ***Saccoglossus*** was described from Madras in 1904 by K. Rammuni. Menon which he named as ***Dolichoglossus bournel***. Van der Horst in 1934 synonymised the genus ***Dolichoglossus*** with the genus ***Saccoglossus Schimkewtsch*** (1892) based on its priority. As no published document was available in favour of the second Indian saccoglossid specimen ***S. madrasensis***reported by Rao (1955), the author refrains from drawing any reference on it in the present discussion. At the same time the author recalls with deep gratitude the pioneer contributions made by the eminent Indian scientist Prof. K. Rammuni Menon on ***Saccoglossus*** (= ***Dolichoglossus***) bournel.

Menon's description on **S. bournel** was based on only two juvenile female specimens dredged from the coastal waters off Madras. The description is incomplete so far the proboscis complex and proboscis stalk anatomy are concerned. The ***Saccoglossus*** specimen from Sundarbans mudflats bears certain apparent resemblances with ***S. bournei***. Proboscis in both the species are elongated and have dorsal and ventral median grooves, though not so pronounced and conspicuous in ***S. bournei***. In both the materials longitudinal muscles of the proboscis are not oriented in concentric rings, instead, the thick muscle fibres display remarkably straight course. Gonads in both cases begin from the first branchial sector of the trunk. Distal tip of the buccal diverticulum or stomochord is bent dorsally in both the cases.

But the dissimilarities between the two species are many and dis tinct. The proboscis coelom in ***S. sundarbanensis***is characterised by a pair of tubular pouches separated by a distinct muscular partition, and extending almost up to the distal end of the proboscis. In the posterior part, the proboscis coeloms accommodate the proboscis complex comprising the buccal diverticulum, heart vesicle, central blood sinus, paired ventrolateral glomerulus and the distal part of the proboscis skeleton. Posterior horns of the proboscis skeleton are inserted on either side of the buccal tube far anterior in the collar instead of the rear end as in **S. bournel**. Collar cord in the present species bears a definite and a well defined neural canal and a neural cord or crest connecting the basement layer of the epidermis. Gonads are lobed in The both sexes, giving rise to two definite genital ridges: no gonoducts In the species under communication.

Similar comparisons with the rest twelve members of the genus illustrate the fact that ***S. sundarbanensis***sp. nov. rightly deserves its separate and exclusive identity as a new member of the genus with special reference to the following specific characters:

(a) Number of gill pores are between 85 to 168 pairs which vary in accordance with the size of the specimen.

(b) Gonads start from the very beginning of the branchial region of the trunk, just behind the collar and form elevated paired genital ridges which in association with the middorsal ridge of the trunk constitute paired branchiogenital grooves in which open the gono pores and gill pores.

(c) The oocytes are nearly spherical and the largest oocytes measuring 200 μm in diameter.

(d) Dorsal proboscis groove is conspicuous and extends almost up to the distal end: ventral groove is less developed and extends upto the posterior quarter of the proboscis.

(e) The most specific character of the species is the proboscis coelom which is represented by paired coelomic pouches extending into the distal end of the proboscis far beyond the proboscis complex.

(f) The longitudinal muscles of the proboscis are never arranged in concentric rings instead, uniformly disposed posteroanteriorly.

(g) The epidermis of the collar has six distinct zones.

(h) The stomochord is very conspicuous with spaceous lumen and varied internal endodermal epithelial configuration; it does not have any dorsal diverticulum.

(i) Glomerulus is a distinct paired structure, one on either side of the central sinus and heart vesicle; no dorsal glomerulus.

Occurrence: Estuarine intertidal mudflats of deltaic Sundarbans of HooghlyMatla estuarine complex, South 24Parganas, West Bengal, India,3 to 4 specimens per square meter occur in average 10 to 100 cm deep in the muddy substratum.

Type materials:

Holotype: S/S No.: 1 is designated to the specimen collected from Prentice Island, Sundarbans, South 24Parganas, West Bengal, India. Collected by B. Singh and A. Choudhury on 15 November 1981; unregistered. at S.D. Marine Biological Research Institute, Sagar Island, India.

Paratype:18 examples, S/S No. 1/218, collection data same as the holotype.

The specific part of the name ***Saccoglossus sundarbanensis*** has been coined after the noted geographical locale, Sundarbans, the most conspicuous mangrove environment of the tropics in the northern hemisphere.

In this discussion part of the dissertation the author intends to elucidate the nature and the possible evolutionary status of the proboscis coelom in enteropneust hemichordates. The observational evidences provided by the macro and microanatomy of the saccoglossid species communicated in the present treatise excite the author to endorse Schepotieff's (1908) contention that the proboscis coelom or protocoel was primitively paired in enteropneusts, which might have persisted in one of the enteropneust species so far discovered from the remote mangrove environment of the deltaic Sundarbans. The series of histological preparations of the proboscis of ***S. sundarbanensis***demonstrate that the proboscis coelom or protocoel in this species is represented as paired 'coels or pouches separated throughout its length, even behind the proboscis complex in the stalk region, by the association of the peritoneal membranes of the paired proboscis coelom (Plate 4B; Plate 5D & E: Plate 6, A, B &C). The author also tempts to maintain the contention that it is the left coelomic pouch of the ancestral paired protocoel which might have persisted in most of the enteropneust hemichordate worms as the adult proboscis coelom and communicated to the exterior by the left dorsal proboscis pore, and the right coelomic pouch or right half of the protocoel having been transformed and translated into the heart vesicle and a related structure rendering combined vascular and excretory functions (Hyman, 1959). This contention is ofcourse subjected to confirmation through pre and postlarval embryo genesis.

Evolution of functional morphology in response to environmental interaction:

The functional part of the anatomical details of the hemichordate worm under communication evokes the worm's most essential biological manifestations in response to its very specialised and vibrating environment. Habitat details and ecology of the species have been dealt with in the second chapter but every part of the worm's morphology, macro or micro, external or internal, demonstrates the very fragile animal's square effort to achieve all possible anatomical excellence only to interact efficiently with its everchanging environment, abiotic or biotic.

The tegumental ciliation all along the tripartite body of the animal associated with a variety of glandular cells, epidermal in origin, serve the dual purpose of foodcord manipulating and coating the burrow with mucous. The second anatomical excellence of this hemichordate worm is the 'proboscis complex' which has not been evolved in any other group of coelomate deuterostomes which are actively burrowing in habit. in mud or sand.

The author has made an honest attempt to elucidate with elaborate illustrations (Plates 9. 10. 11. 12) both the mechanical and physiological role of this proboscis complex where most of the vital organs of the animal have been concentrated. It is the proboscis coelom (protocoel) with paired pouches in the present species, shared the most important responsibilities for the formation, and later accommodation, of the proboscis complex in concert with the larval ectoderm. The later contributed to the formation of the ciliated epidermis and epidermal neural layer and the chitenoid proboscis skeleton that supports the complex onits dorsal and proximal end. Except the buccal diverticulum, other components of the complex viz.. the central sinus, cardiac vesicle, dorsal mesentery, paired lateral glomeruli and the intricate musculature of the proboscis, are all derived from the embryonic coelomic epithelium.

In this vermiform sluggish burrowing animal. proboscis appears to be the most active part engaged in continuous tunneling to explore the food environment and hence, the maximum energy spent up during Its twenty four hours activities, is possibly through the proboscis segment. This segment is equipped with enormous amount of longitudinal muscle fibres justifying its functional responsibilities. These muscle fibres originate from the basement membrane of the epidermis which Is the reminiscent of the parietal peritomeum (somatic mesodermal epi thelium) of the proboscis coelom. These muscle fibres are inserted at the basal part of the proboscis and proboscis complex including the chondroid skeletal arches sheltering various blood sinuses.

Muscles are the main sites of energy expenditure for their continuous extension and contraction phases during the incessant working and reworking for explorative burrows. To replenish their spent up energy they are copiously supplied with blood vessels or capillary spaces which in turn get supply from the centrally located blood sinus of the proboscis complex. During this active metabolic process considerable amount of nitrogenous wastes are supposed to be generated, which naturally, need to be got rid of and for this the unique excretory devise, the glomerulus evolved. This intricate vascularised organ is formed by repeated epithelial evaginations from the lateral sides of the central sinus, and being projected into the respective coelomic spaces of the proboscis, is supposed to remove the generated nitrogenous wastes. In this excretory process the nephrogenous cells or chloragogen cells budded off from the glomerular epithelium into the coelomic spaces might have some role to play. No experimental confirmation in favour of this contention is yet available and hence demands thorough investigation. In comparison to the other members of the genus ***Saccoglossus*** the glomeruli are well developed in the present species.

The other important component of the proboscis complex is the buccal diverticulum which was misconceived as notochord by the earlier authors of hemichordates. Hyman (1959) vehemently disfavoured this contention and categorically dismissed the argument that the buccal diverticulum ever possessed any remote relationship or similarities with the embryogenesis or functional anatomy of notochord of the chordates. It is revealed through the meticulous microanatomical studies in the present species that buccal diverticulum displays extraordinary histological configurations, the functional implications of which are yet to be ascertained. It has been portrayed in the microanatomy chapter of the dissertation that the buccal diverticulum, throughout its length. showed contrasting regional endodermal epithelial modifications. Through meticulous histochemical examinations only the physiological responsibility of these regions could be established. The present author envisages to deploy a separate research project on the fine structure and histochemistry of the entire length of the buccal diverticulum of ***S. sundarbanensis*** sp. nov. to unvail the possible role of this conspicuous component of the proboscis complex and its possible physiological role which may signify it's so prominent occupation in hemichordate anatomy.

Not much embryological work has been done to elucidate the transformation of the larval protocoel to adult hemichordate worms except that of BurdonJones (1952) on ***Saccoglossus*** horstl. Circumstances did not permit the present author to carry out the embryological studies of the ***saccoglossid*** species during the present dissertation work but it has been envisaged to take up a second phase of study on the animal to highlight two important phenomena:

(1) Functional role of the entire length of the buccal diverticulum (through histochemical and transmission electron microscopical (TEM) study).

(2) Detail embryogenesis of protocoel or proboscis coelom.

Pending this investigation, the author refrains from proposing any amendment of the existing diagnostic features of the genus ***Saccoglossus*** nor proposing any new genus under the family Harrimaniidae to provide shelter for the new species ***Saccoglossus sundarbanensis*** because of having double coelomic pouches in the proboscis.

PLATE 4

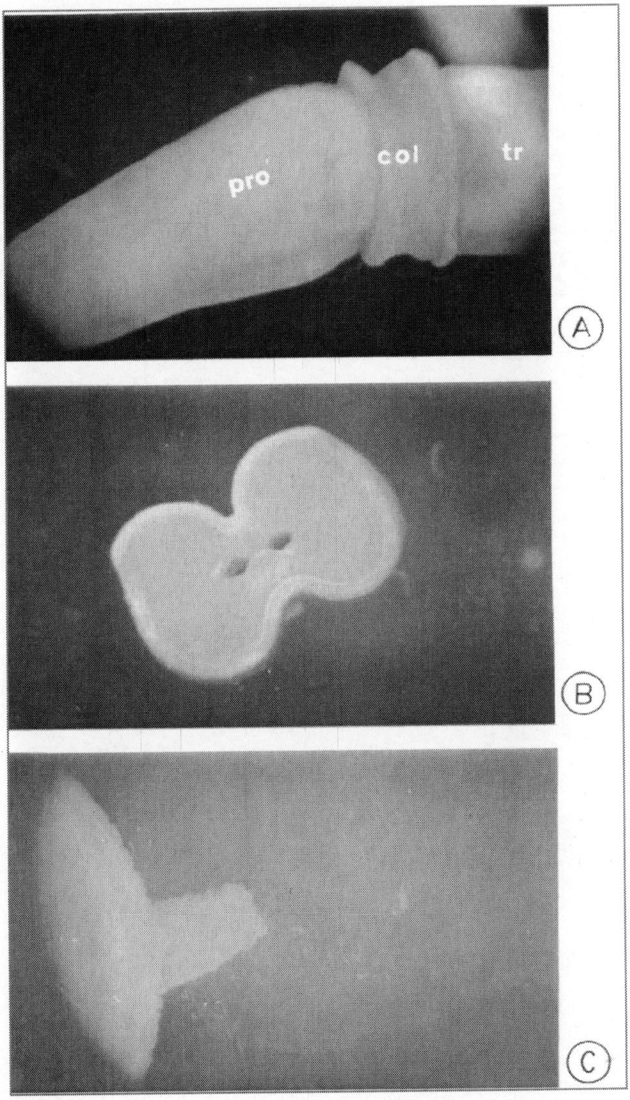

(A) Proboscis (pro), Collar (col) and partial Trunk (tr) region of ***Saccoglossus sundarbanensis*** sp. nov.

(B) Cross section of Proboscis (= anterior part) showing double protocoel (pc) and the partition wall (st).

(C) Anal region of ***S. sundarbanensis***sp. nov.

PLATE 5

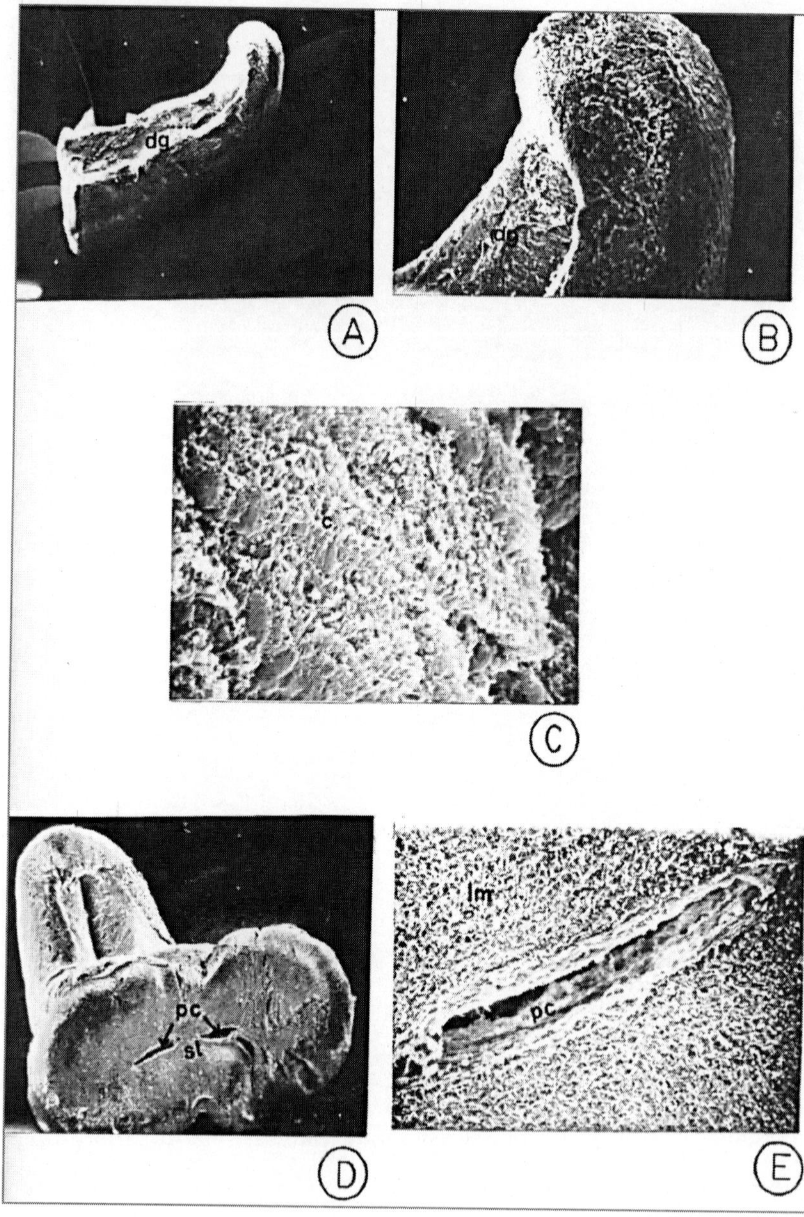

Scanning microphotograph of the proboscis parts showing :

(A)Proximal proboscis part showing dorsal groove (dg). [X 12]

(B)Anterior tip of the proboscis more magnified. [X 200]

(C)Surface coat of the proboscis showing ciliation (c). [X 1600]

(D)Paired proboscis coelom (pc) and its partition wall (st). [X 25]

(E)Right proboscis coelom (pc) and longitudinalmuscles (im).[X 200]

PLATE 6

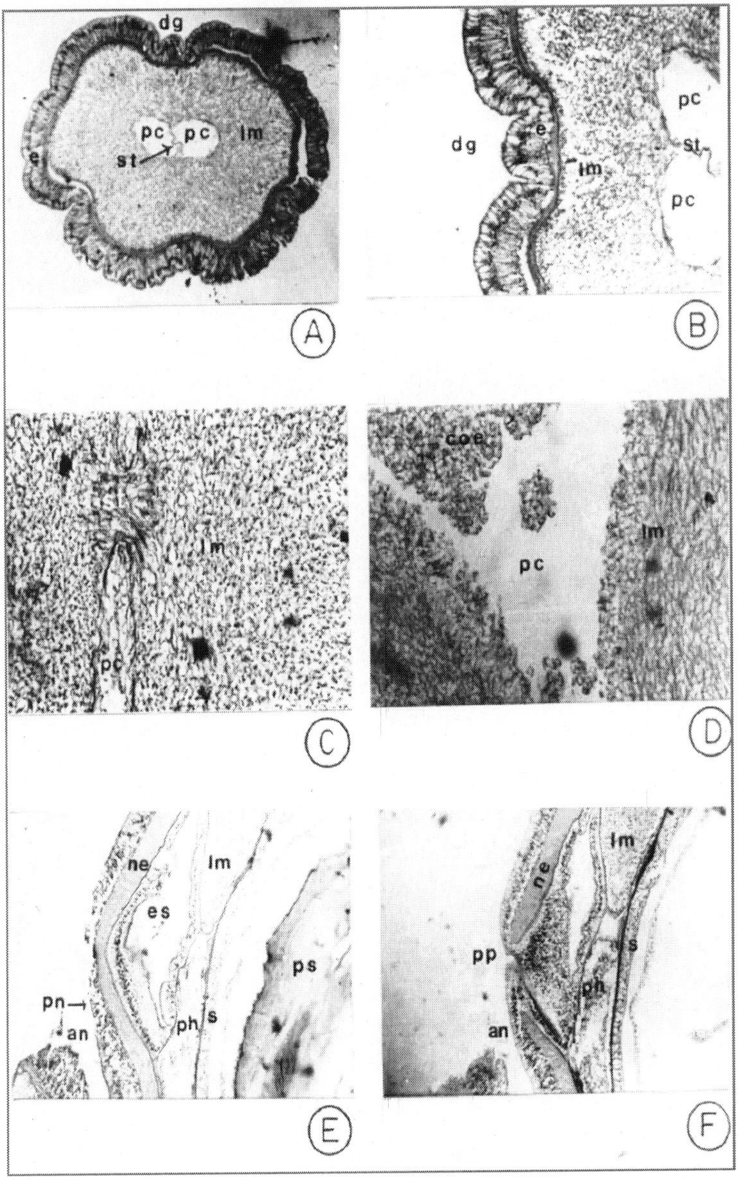

(A)Transverse section through anterior tip of proboscis showing double proboscis coelom or protocoel (pc) with thin septum (st), longitudinal muscles (1m), epidermis (e) and dorsal groove (dg). [10 X 6 μm]

(B)Same as 'A' but magnifiedpart of proboscis. Dorsal groove (dg) in the anterior part of the proboscis [10 X 45 μm]

(C)Muscular partition (st) of the proboscis coelom (pc) in the middle of the proboscis. [10 X 40 X 0.8 μm]

(D)Coelomocytes (coe) in the proboscis coelom (pc). [10 x 45 μm]

(E) Longitudinal section through proboscis neck (pn) showing end sac (es), nerve layer (ne), anterior neuropore (an), and proboscis skeleton (ps). [10 X 10 X 1.25 μm).

(F)Longitudinal section through proboscis neck showing proboscis pore (pp) and its chamber. Note the extension of the perihaemal space (ph) from the trunk and longitudinal muscles (Im) of the proboscis below which lies the stomochord (s). [10 X 10 X 1.25 μm]

PLATE 7

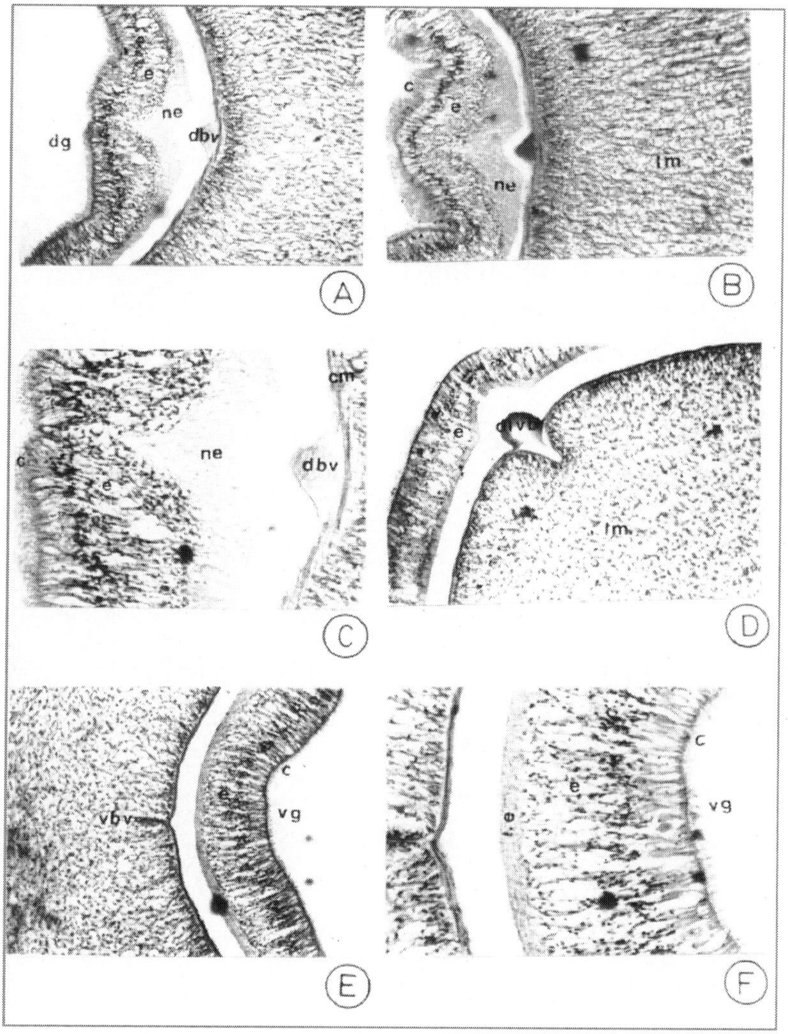

Transverse sections through middle part of the proboscis showing

(A) Dorsal groove (dg), epidermis (e) with coat of cilia (c) nerve layer (ne). dorsal blood vessel (dbv) and longitudinal muscles (lm). [10 X 10 X 1.25 μm]

(B) Same as 'A' but note the thickening of nerve layer (ne) below ciliated epidermis (e) of dorsal groove (dg). [10 X 40 X 1.25 μm]

(C) Ciliated epidermis (e) magnified. [10 x 10 x 1.25 μm]

(D) Dorsolateral blood vessel (dlbv) of the proboscis and thick layer of longitudinal muscles (lm). (10 X 10 X 1.25 μm]

(E) Ventral blood vessel (vbv), uniform ventral thin nerve layer (ne), epidermis (e) above ventral groove (vg).[10 X 10 X 1.25 m]

(F) Epidermis (e) and its cilia in the ventral groove of proboscismagnified. [10 X 100 X 1.25 μm]

PLATE 8

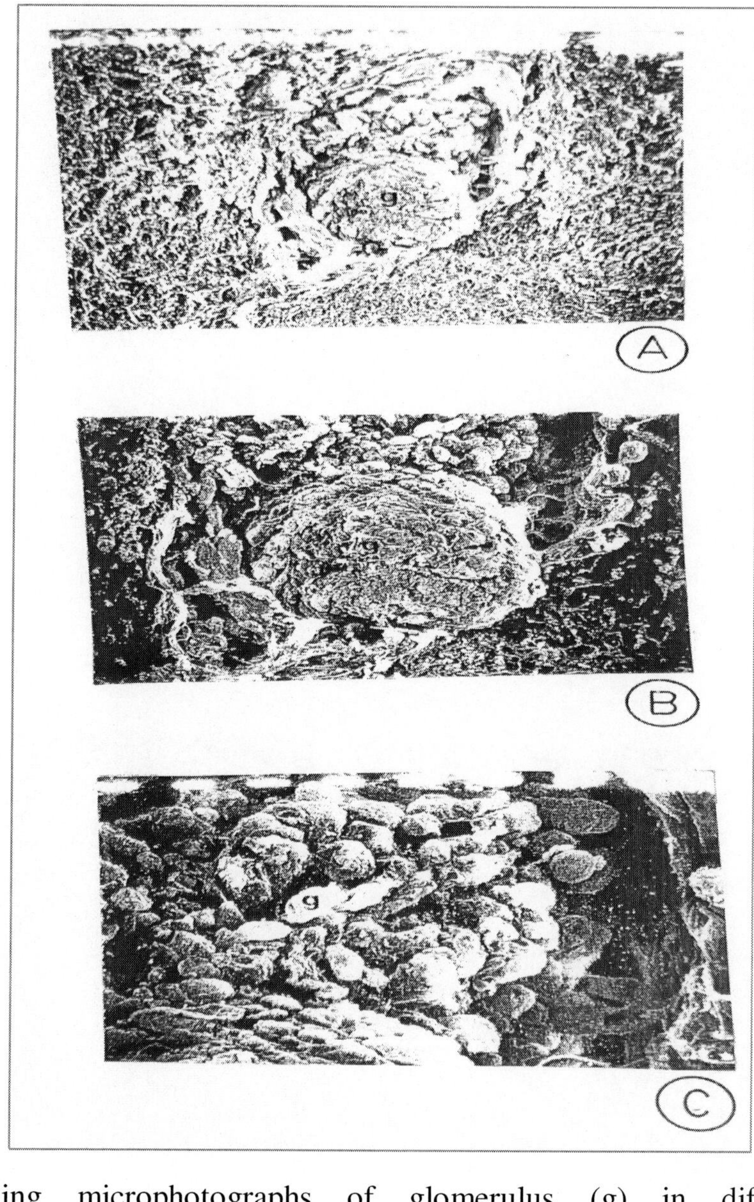

Scanning microphotographs of glomerulus (g) in different magnifications.

[AX 200, BX 400, CX 800]

PLATE 9

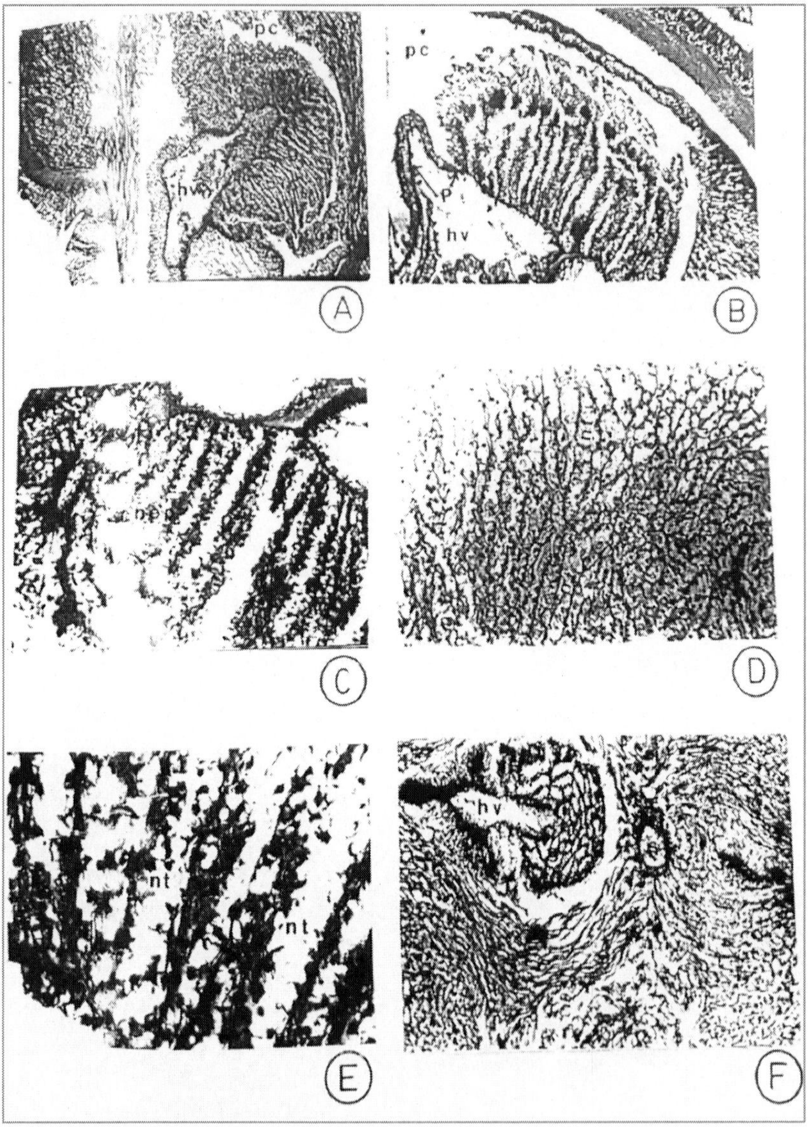

Microphotographs of glomerulus (g).

(A) Longitudinal section through glomerulus (g) showing its position in the basal part of the proboscis coelom (pc) and its relation with heart vesicle (hv). [4 X 10 X 1.25 μm]

(B) Coelomic epithelium (p) ramifying to form glomerulus and a heart vesicle (hv).[10 X 10 X 1.25 μm]

(C)Same as 'B' and showing nephrocytes (nep) or chloragogencells.[10 X 40 X 1.25 μm]

(D) Longitudinal section through glomerulus showing compact nephrictubules (nt) in the central part of the glomerulus. [10 X 10 μm]

(E) Nephric tubules (nt) magnified. [10 X 100 X 1.25 μm)

(F) Glomerulus below the tip of the stomochord (s). Heart vesicle (hv) is connected with blood vessels (bv). [10 X 10 X 0.8 μm]

PLATE 10

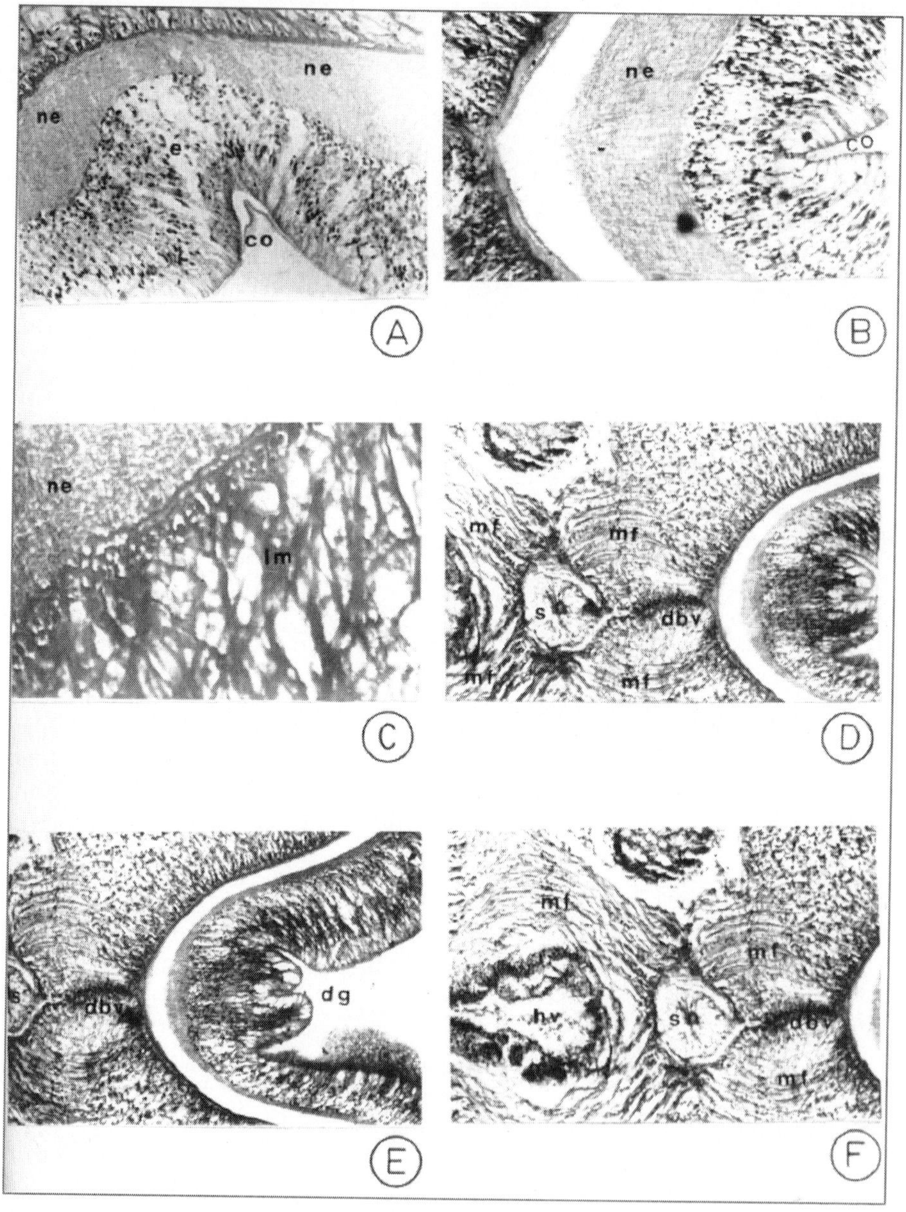

Sections through the proboscis base.

(A) Ventrolateral groove of the ciliary organ (co) with thick basal epidermis (e) and nerve layer (ne). [10 X 10 x 1.25 μm]

(B) Deep indentation of the ventral groove of the ciliary organ (co) at the proboscis base and very thick nerve layer (ne).[40 X 10 X 1.25 m]

(C) Anastomosing longitudinal muscle fibres (Im) of the proboscis.[10 X 100 X 1.25 µm]

(D) Muscle fibres (mf) encircling the dorsal blood vessel (dbv) and stomochord (s).[10 X 40 X 0.8 m]

(E) Dorsal blood vessel (dbv) below dorsal groove (dg).[10 X 40 X 0.8 m]

(F) Muscle fibres (mf) as in 'D' encircling heart vesicle (hv). stomochord (s) and dorsal blood vessel (dv). (10 X 40 X 0.8 m)

PLATE 11

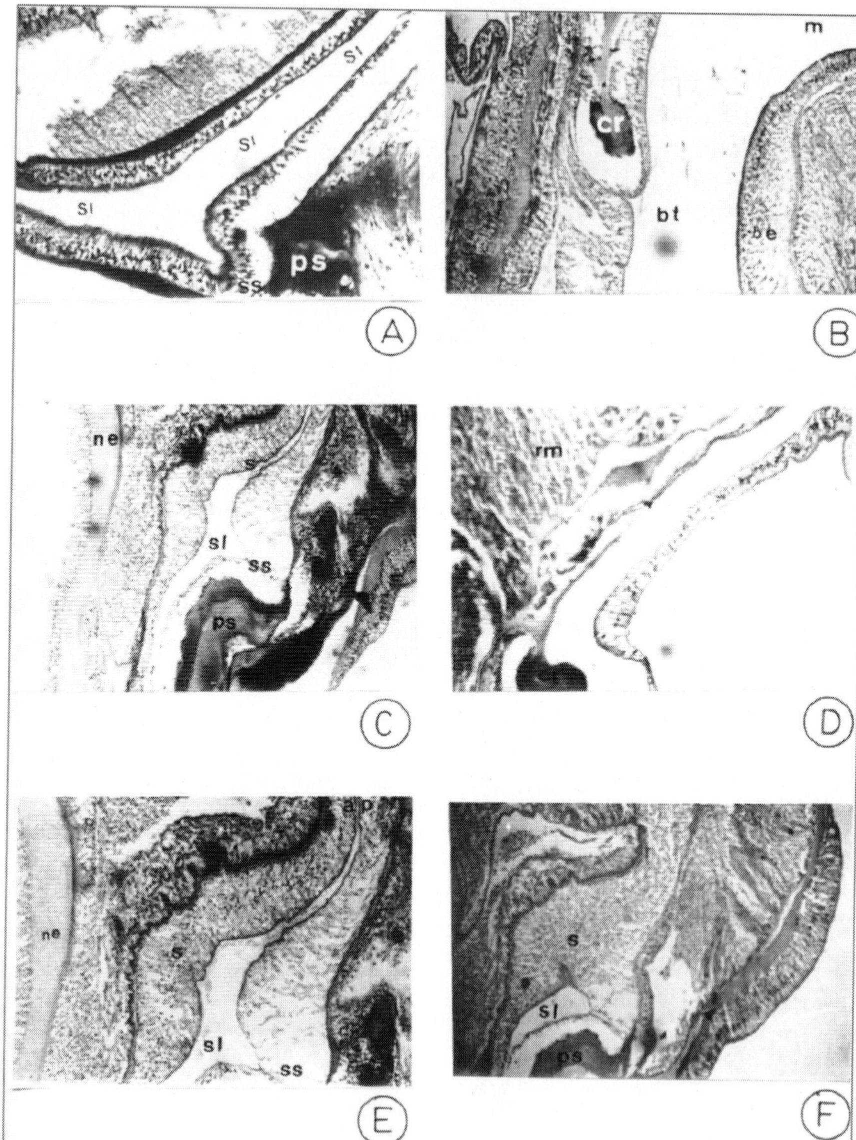

Longitudinal sections through collar and proboscis base showing stomochord :

(A) Longitudinal section of stomochord (s) showing stomochord lumen (sl). [10 X 40 X 1.25 μm)

(B) Cornue (cr) of proboscis skeleton (ps) embedded in dorsalroof of the buccal cavity (bt).[10 x 10 µm]

(C) Stomochord (s) resting on proboscis skeleton (ps) and showing its ventral Sshaped extension (ss).[10 x 10 x 0.8 µm]

(D) Retractor muscles (rm) attached with proboscis skeleton (ps) in collar.

(E) Stomochord with its widest lumen (sl) and upward bend of its apex (ap). [40 X 10 X 1.25 µm]

(F) Same as (E)

PLATE 12

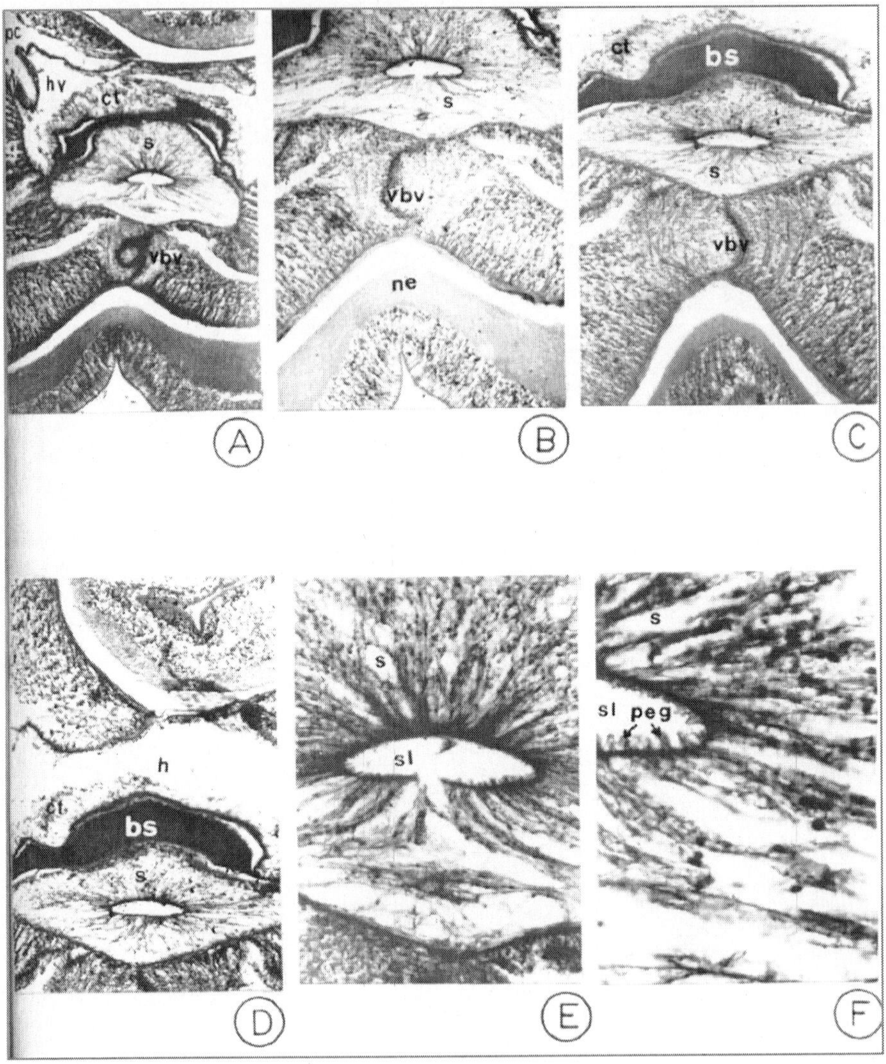

Transverse section through stomochord and blood sinus in the proboscis base.

(A) 'Proboscis complex' [= stomochord (s). blood sinus (bs). heart vesicle (hv) and glomerulus (g)] in the lumen of the proboscis coelom (pc). [10 x 10 x 1.25 μm]

(B) Ventral blood vessel (vbv) connecting the stomochord with nerve layer (ne) just above the ventral groove (vg). [10 x 10 x 1.25 μm]

(C) Cellular thickening (ct) at the basal epithelium of the heart vesicle (hv) above the wide lumen of the 'blood sinus' (bs). [10 X 40 X 1.25 μm]

(D)Cellular thickening (ct) at the basal epithelium of the heart vesicle (hv) above the wide lumen of the blood sinus (bs).[10 X 40 X 1.25 μm]

(E) Stomochord (s) displaying the 'peg' like dentition (pg) projected in the stomochord lumen (sl) in the basal zone of the proboscis. [10 X 40 X 0.8 μm]

(F) Magnified view of 'Peg' like dentition projected in the stomochord lumen (sl).[10 x 100 x 1.25 μm]

PLATE 13

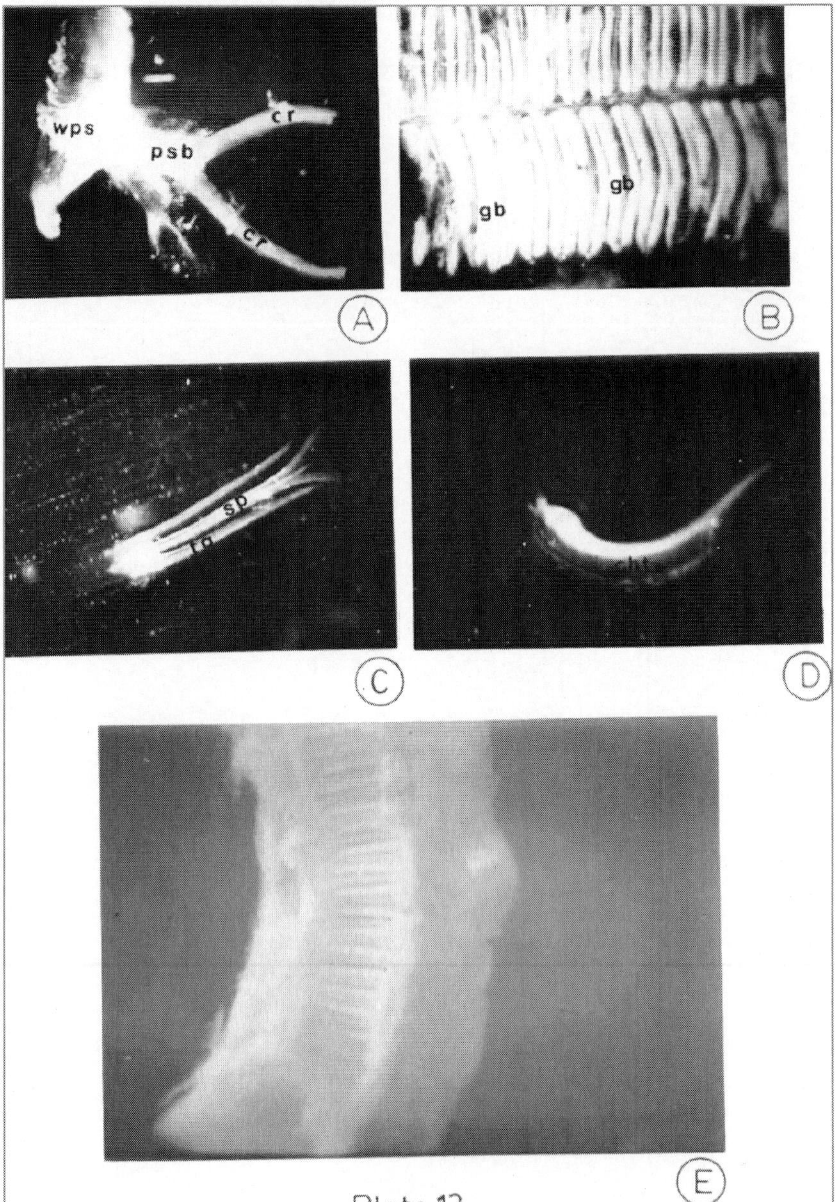

Plate 13

(A) Dissected proboscis skeleton showing its Y-shaped structure with wide base of proboscis skeleton (wps). body of the proboscis skeleton (psb) and horns (cr). [15 X 20]

(B) Branchial chamber cut open dorsally to show the gill bars (gb). [15 X 20]

(C) Dorsal view of the gillskeleton showing forked septa (sp)and unforked tongue bars (tg).[15 X 20]

(D) Side view of the gillskeleton showing its dorsolateral chitinous flap (cht). [15 X 20]

(E) Pharynx cut dorsoventrally showing U-shaped perforation of the pharyngeal wall and gill bars. [15 X 20]

PLATE 14

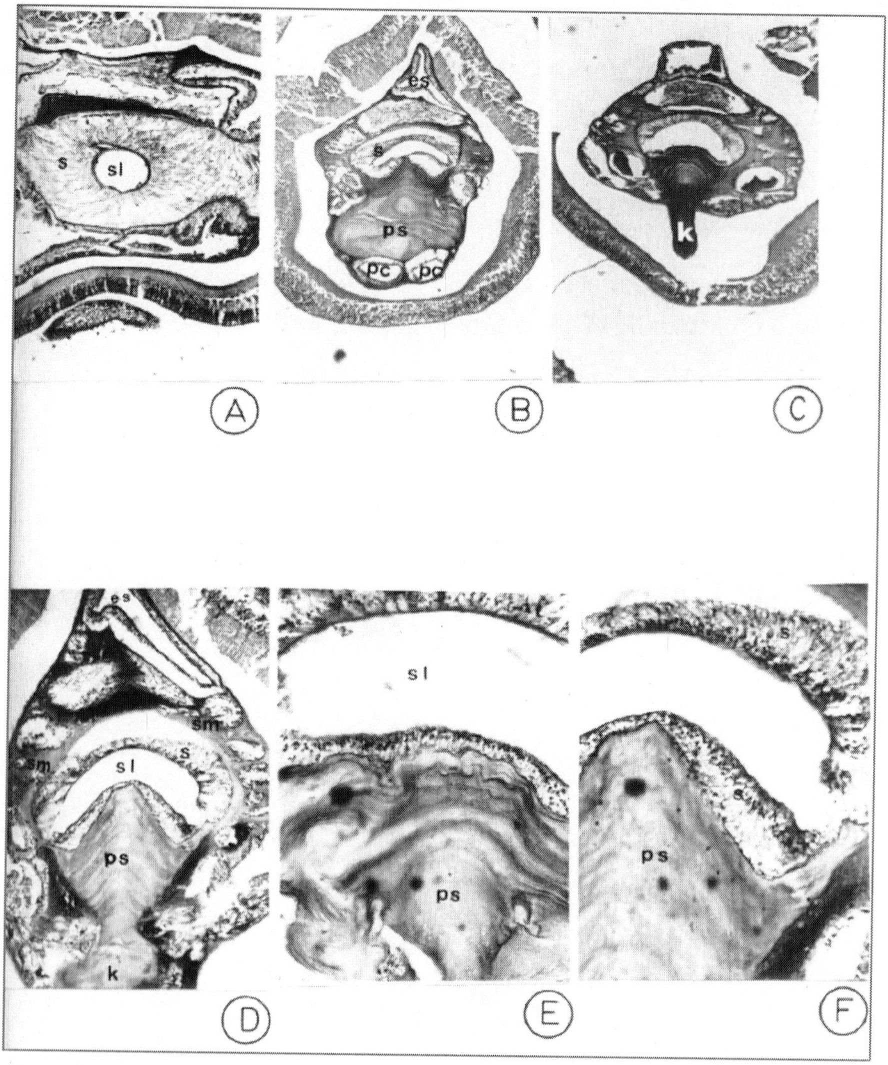

Transverse sections through proboscis neck.

(A) Stomochord lumen (s1) which becomes circular in shape at the junction of proboscis and broboscis stalk. (4 X 10 X 0.8 μm)

(B) End sac (es), stomochord (s) and coelomic spaces surrounded by proboscis skeleton (ps) and note the two protocoel cavities below the skeleton (pc). (4 X 10 X 0.8 μm)

(C) Keel (k) of the proboscis skeleton. (4 X 10 X 0.8 μm)

(D) Chitinous layers deposited in the proboscis skeleton (ps) and the semicircular rings (sm) completely enveloping the stomochord. (10 X 40 X 0.8 μm)

(E) The central portion of the proboscis skeleton in the anterior region of the collar magnified. [10 x 100 x 1.25 μm]

(F) Note the semilunar lumen of the stomachord (s1). the cells of the dorsal lining membrane are taller than the cells of the ventral membrane of the stomochord lumen: cell nuclei are always apical in position. [10 X 10 X 1.25 μm]

PLATE 15

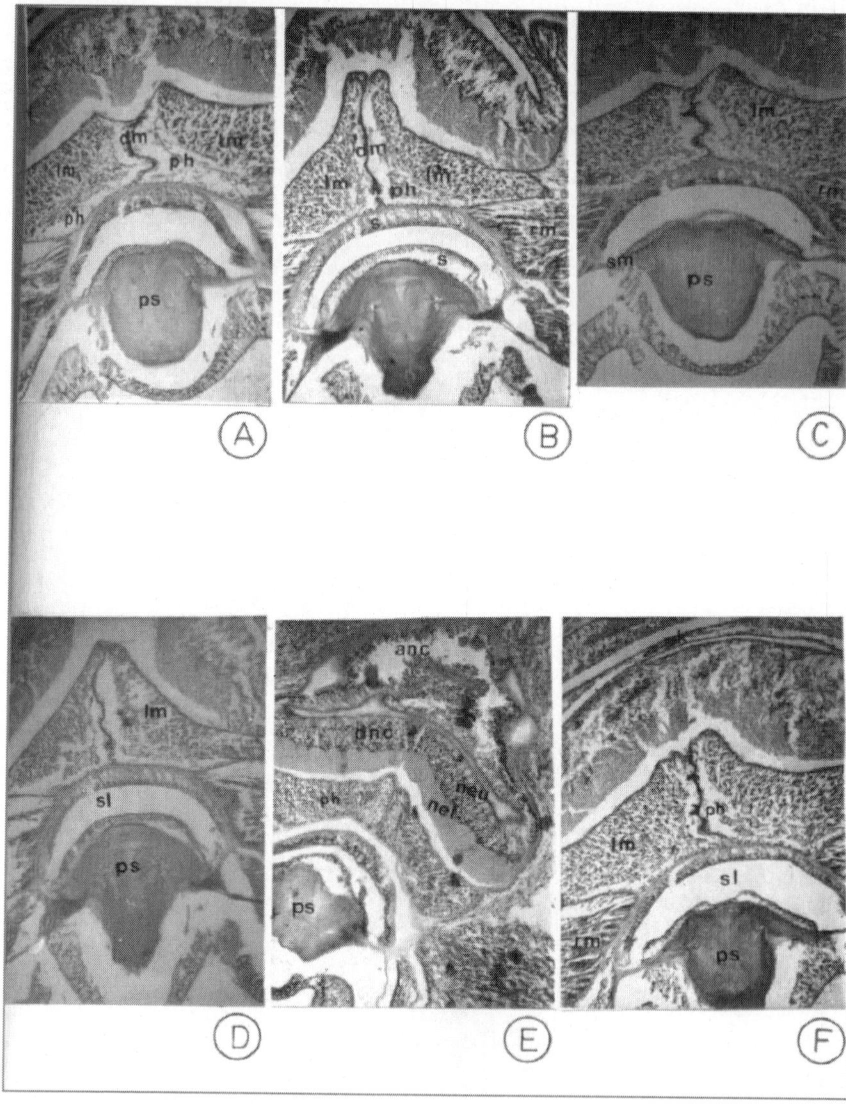

Transverse section through collar:

(A) Perihaemal spaces (ph) separated by dorsal mesentery (dm) and occupied by longitudinal muscles (1m) of the trunk.[10 x 10 x 1.25 μm]

(B) The dorsal and ventral cellular walls of the stomochord are of uniform height. [10 X 10 X 1.25 μm]

(C) From the lateral sides of the proboscis skeleton (ps), skeletal plates arch out forming semicircular ring (sm) completely enveloping and supporting the stomachord and on it retractor muscles (rm) are anchored.[10 X 10 X 1.25 μm]

(D) Same as 'C'.

(E) Extension of anterior neural canal (anc) upto the hollow dorsal nerve cord (nc). Note the dorsal 'neuron layer' (neu) and ventral 'nerve fibre' (nef) layers of the collar nerve cord (dnc). [10 x 10 x 1.25 μm]

(F) Same as 'C' but neural keel (nk) is present.

PLATE 16

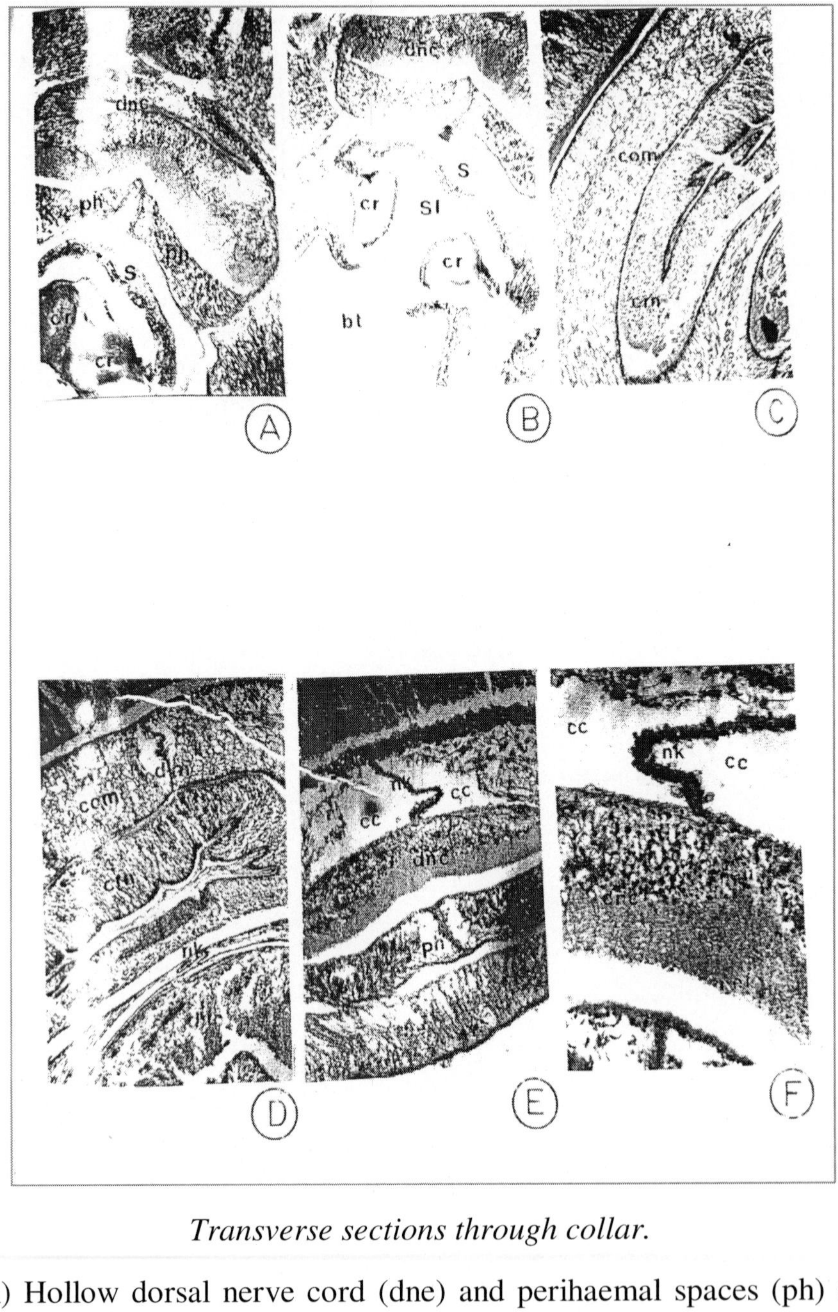

Transverse sections through collar.

(A) Hollow dorsal nerve cord (dne) and perihaemal spaces (ph) are dorsoventrally compressed. Ventral wall of the stomochard (s) is membranous where the two horns (cr) of the proboscisskeleton(ps) meet together with the main body.[10 X 10 X 1.25 μm]

(B) The stomochord lumen (sl) opens into the buccal cavity (bt) in between two horns (cr) of the proboscis skeleton.[10 x 10 x 1.25 μm]

(C) Crescentic notch (crn) formed by the invaginated dorsal collar flap surrounded by collar muscles (com). [10 x 40 x 0.8 μm]

(D)Same as 'C' but dorsal mesentery (dm) of the collar and neural keel (nk) are present.

(E) Neural keel (nk) connecting dorsal nerve cord (dnc) and epidermis of the collar epithelium (ce) acting as a partition wall between two collar coeloms (cc). Perihaemal spaces are dorsoventrally compressed between nerve cord and bucal epithelium (be). [10 X 40 X 1.25 μm]

(F) Neural keel (nk) and nerve cord (nc) magnified. Note the dorsal neuron layer (neu) and ventral nerve fibre layer (nef) of nerve cord (dnc).[10 x 100 x 1.25 μm]

PLATE 17

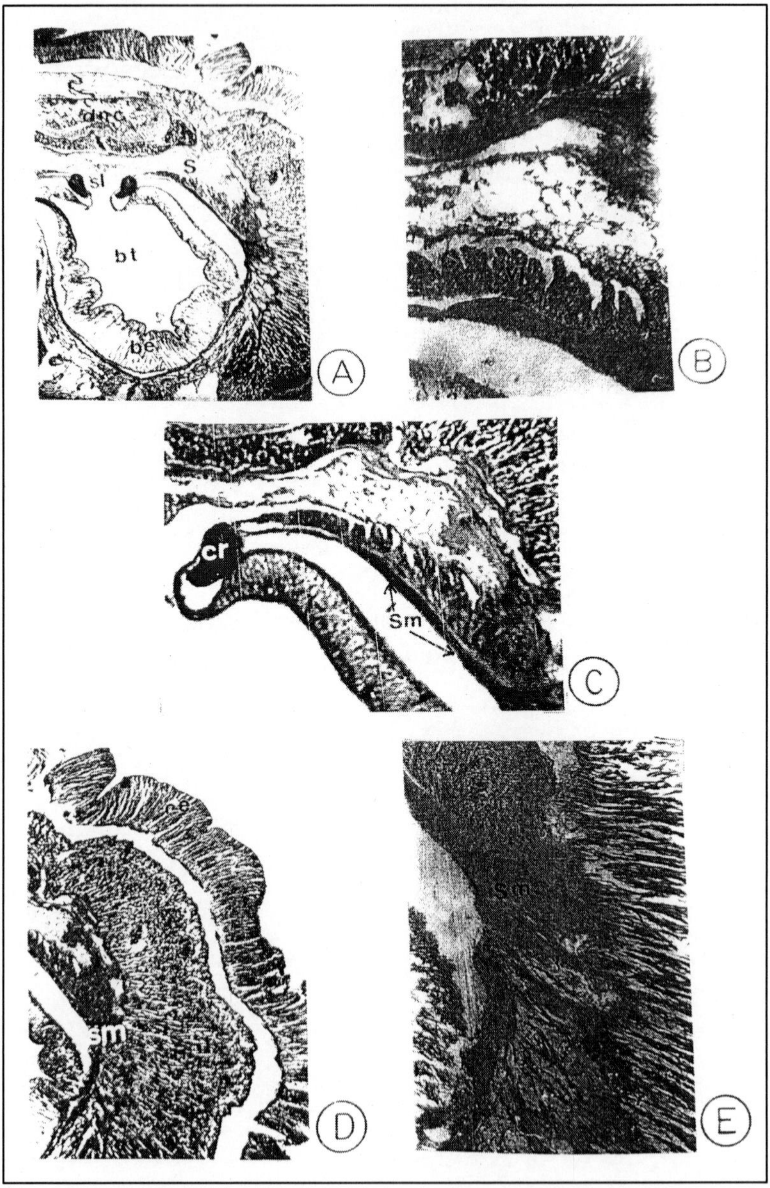

Transverse sections through collar.

(A) Stomochord (s) originating from the dorsal roof of the buccal tube (bt) maintaining direct connection between buccal tube and its lumen (s1). The stomochord is supported by proboscis skeleton (ps) and its cartilaginous rings (sm). Note the dorso-lateral expansion of stomachord.[10 X 10 X 1.25 μm]

(B) 'Villi-like' structures (vl) in the ventral wall of the dorsolateral expansion of the stomochord (s). [10 x 100 x 1.25 μm]

(C) 'Villi-like' structures (vl) touches the roof of the stomochord and are supported by cartilaginous rings (sm). [10 X 45 X 1.25 μm]

(D) Rectractor muscles of the collar (rm) supported by the semicircular cartilaginous rings of the proboscis skeleton (sm). collar epithelium (ce). [10 x 10 x 1.25 μm]

(E) Rectractor muscles (rm) magnified. [10 x 100 x 1.25 μm]

PLATE 18

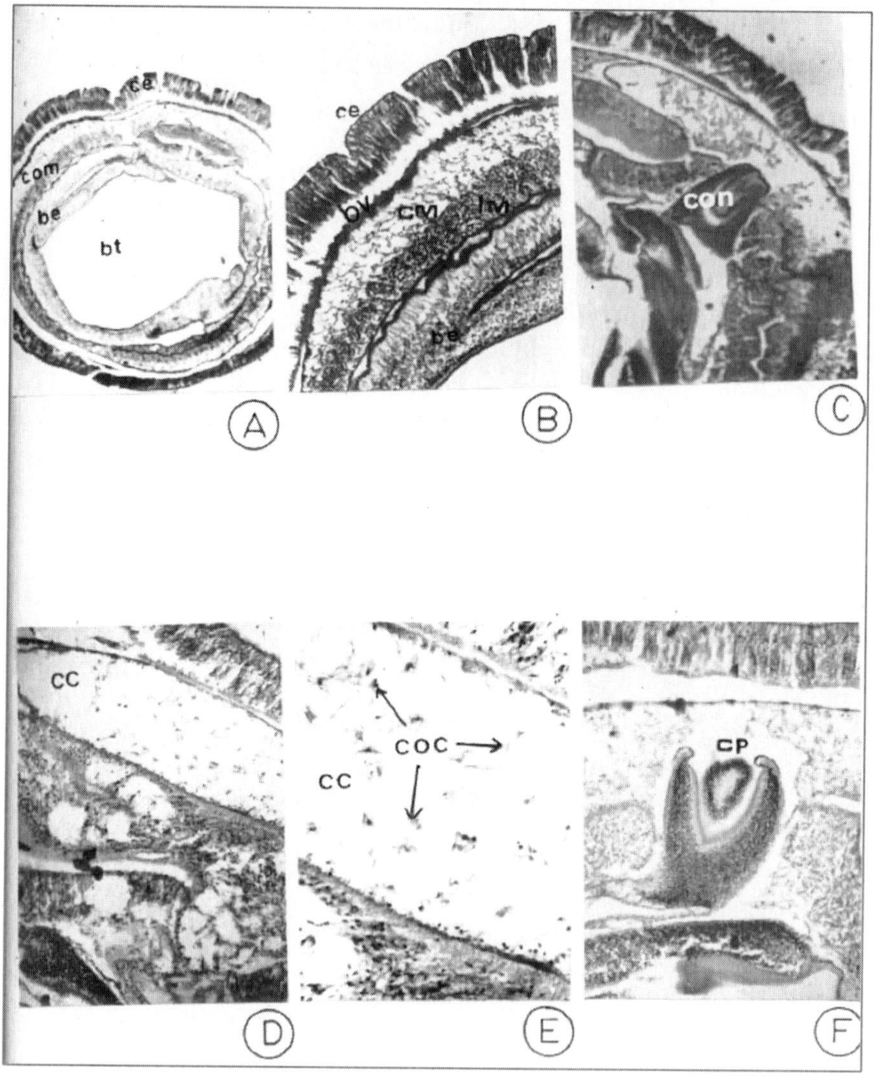

Transverse sections through collar.

(A) Section through collar showing epidermis (ce) blood vessels (bv) collar muscles (com) buccal epithelium (be) and buccal cavity (bt). [4 x 10 x 0.8 μm]

(B) Series of peripheral blood vessels (bv). circular (cm) and longitudinal (Im) collar muscles and buccal epithelium magnified. [10 x 10 x 1.25 µm]

(C) Collar canal (con). [10 X 10 X 0.8 µm]

(D) Collar coelom (cc) filled with coelomocytes (coe). [10 x 10 x 1.25 µm]

(E) Coelomocytes (coe) magnified. [10 x 10 x 1.25 µm]

(F) Collar pore (cp) opening in the first gill pouch of the pharynx. [10 x 10 x 1.25 µm]

PLATE 19

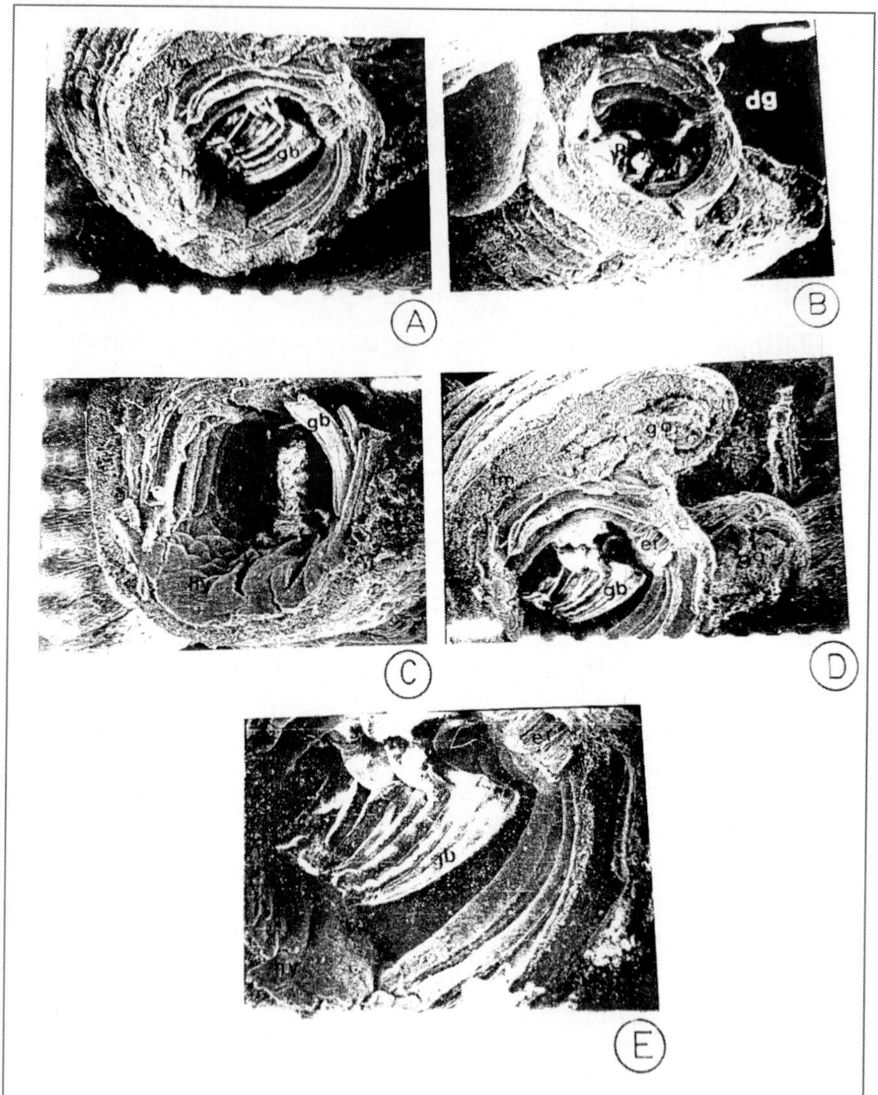

Scanning microphotographs of branchial sector showing:

Longitudinal muscles of the trunk (Im).

Pharynx (phx).

Branchial lamellae or gill bars (gb).

Epibranchial ridge (er).

Hypobranchial ridge (hy).

Dorsal groove of the trunk (dg).

Genital ridge (gr) packed with gonads (go)

A, B, C and D magnified to 50 times and E magnified to 100 times.

PLATE 20

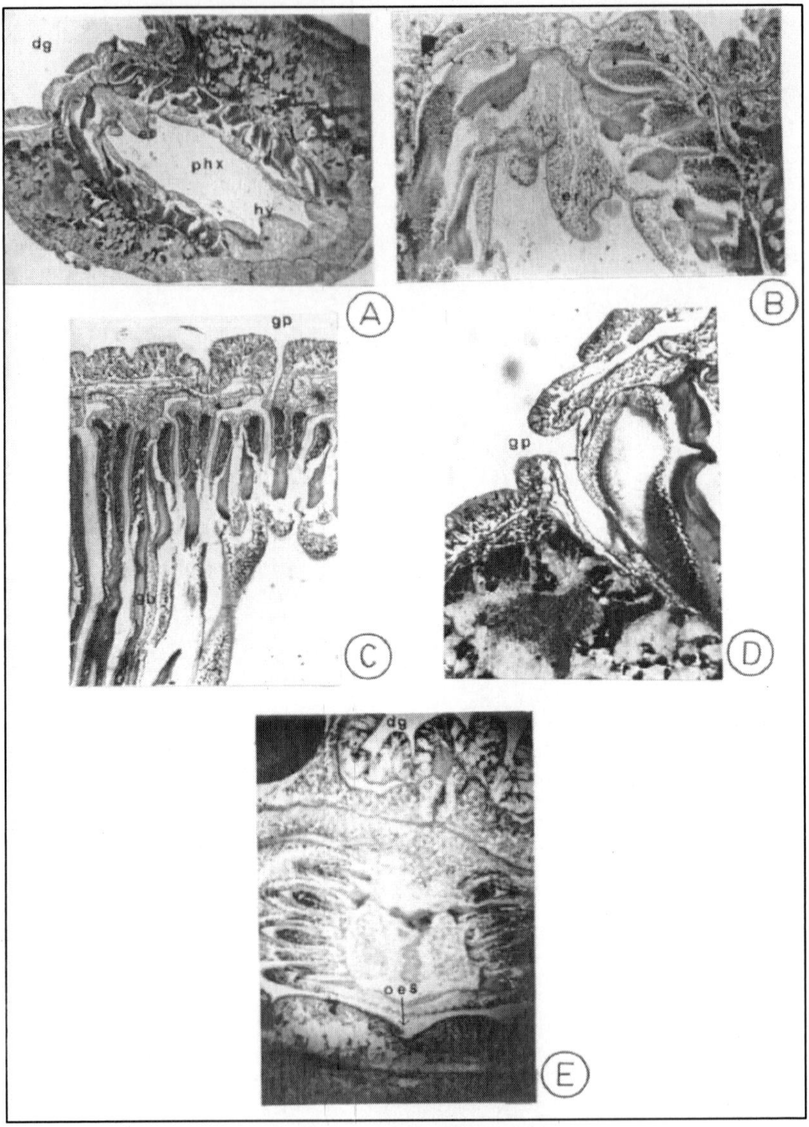

Transverse and longitudinal sections through branchial zone and oesophagus.

(A)Transverse section through middle of the pharynx showing 'epibranchial ridge (er) hypobranchial zone (hy) gill bars (gb) and the pharyngeal chamber (phx). (4 X 10 X 0.8 μm]

(B) Epibranchial ridge (er) magnified. [10 x 40 x 1.25 μm]

(C) Longitudinal section through pharynx showing gillbars (gb) and 1st gill pore (gp). [10 x 10 x 0.8 μm]

(D) Opening of the gillpore (gp). [10 X 40 X 0.8 μm]

(E)Transverse section through oesophagus (oes) showing enlargement of the dorsal wall, which function like a valve. [4 X 10 X 1.25 μm]

PLATE 21

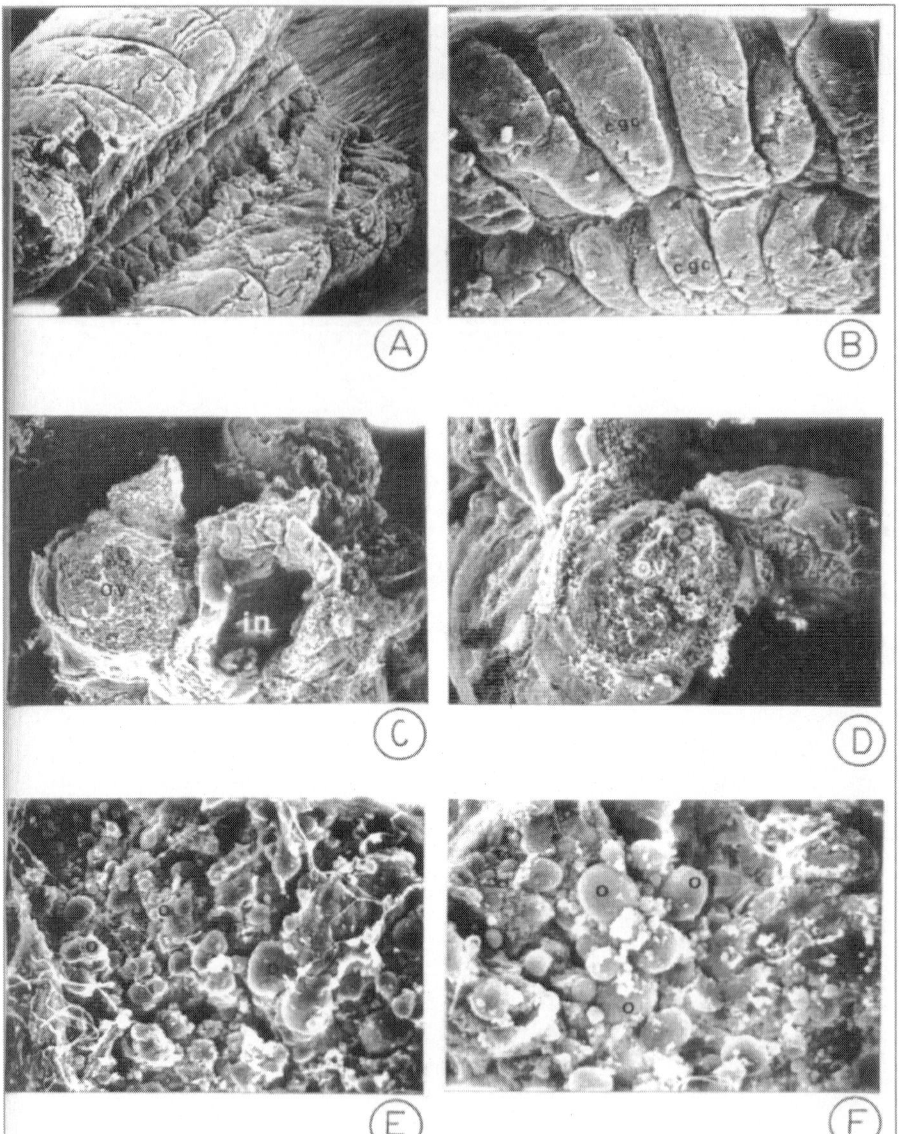

Scanning microphotographs of the trunk and gonads.

(A)Impression of the dorsal nerve cord (dnc) in the dorsal grooveof the trunk.[X 50]

(B) Cluster of gland cells (cgc) in the posterior part of thetrunk. [X 50]

(C)Lumen of the intestine (in) and ovary (ov) on the right side. [X 50]

(D)Ovary (ov). [X 100]

(E) Oocytes (0) in ovary. [X 400]

(F) Oocytes (o) in ovary magnified. [X 800]

PLATE 22

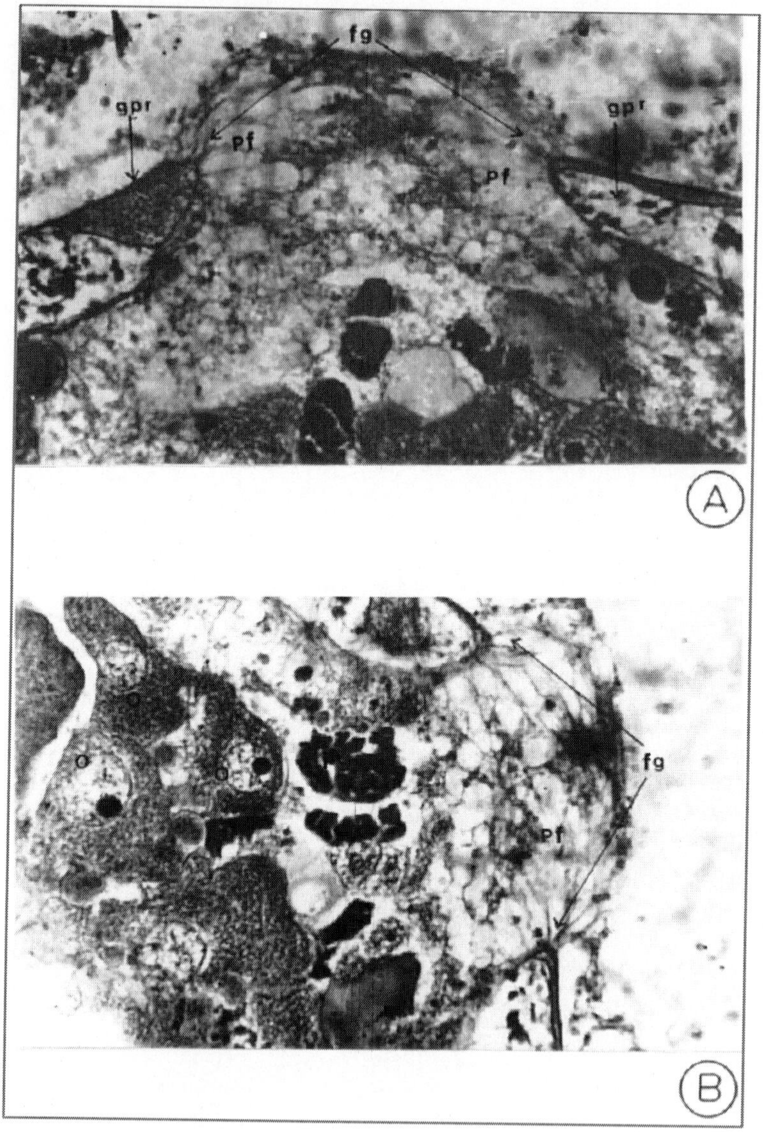

(A) Female gonopore (fg) guarded by plug of fibrilar structures (pf); cut ends of gonopore rim (gpr) are visible. [10 X 100 X 0.08 μm]

(B) Developing oocytes (o) of different stages in ovary (ov): glycogen globules (gg) are still present in ovary.[10 X 100 X 0.8 μm]

PLATE 23

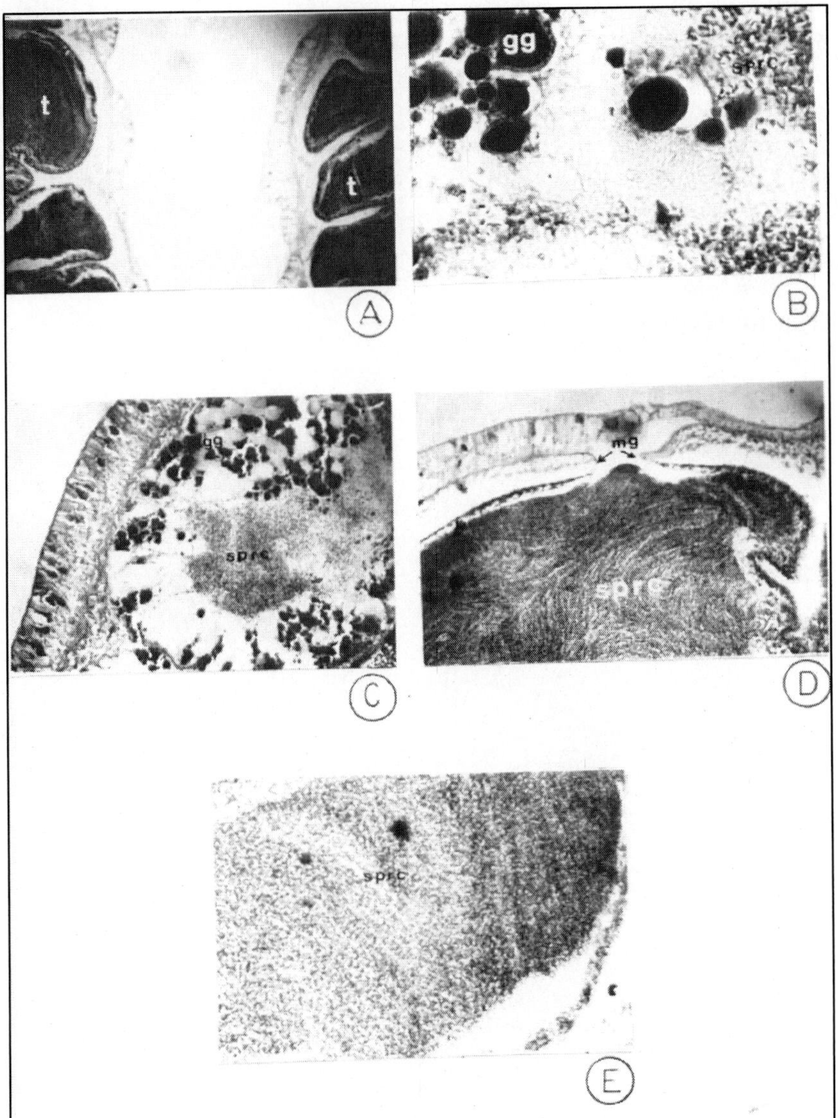

Transverse sections through Immature and fully matured male gonads.

(A) Many lobed male gonads (t) in the genital wing packed withfully matured sperms (spr).[4 X 10 X 1.25 μm]

(B)Section through male gonad packed with developing immaturespermatocytes (sprc) and glycogen globules (gg).[10 x 100 x 0.8 μm]

(C) Testis packed with sperms (spr) and glycogen globules (gg). [10 X 40 X 0.8 μm]

(D) Testis packed with fully matured sperms (spr). Note the site of male gonopore (mg). [10 x 10 x 1.25 μm]

(E) Mature sperms, in testis. [10 x 40 x 0.8 μm)

CHAPTER II

INTRODUCTION

The dissertation highlights in Chapter11 the ecology, ethology and biology of ***Saccoglossus sundarbanensis*** sp. nov. It also deals with the ecology of estuarine area identified for the present study. and the luxuriant benthic community residing here along with the hemichordate worm as one of its important member. Physiography and climate of the HooghlyMatla estuarine complex have already been dealt with in the first chapter (page 9 to 12) of the dissertation.

Sand and mud flats are the most extensive intertidal habitats in the majority of the tropical estuarine environments, particularly the mangrove climaxes of Indian estuaries. Within the estuaries these tidal flat habitats may occur all the way along the salinity gradient. Beaches and tidal flats are affected by both vertical and horizontal gradients. Factors varying vertically include duration of emergence and submergence, influence of weather, and sediment composition. Horizontal gradients, more important for tidal mudflats, run from the open shore into the estuary and include salinity, wave exposure, sediment composition and tidal amplitude.

A closer examination of the tidal mudflats of deltaic Sundarbans shows the signs of a galaxy of life that exists beneath and above its surface, the evidence of which is laid down by their highly dramatic activities like biotarbation and a mess of changing contours caused by various organisms. The larger animals that leave these tell-tale patterns are usually called macrofauna or macrobenthos but these comprise only a part of a fauna of the sediments. Living between the sand grains, or on muddy beaches within the mud. is a whole variety of small animals that will pass through the fine meshes of the screen. These small animals are called the interstitial fauna, since they live in the interstices between sand grains or clay particles, but the general name for all these animals are the meiofauna and microfauna depending on their size range.

Estuarine mudflats are subjected to two overriding physical constraints which regulate the growth of the biota they can support these are silt and saline water. By tradition, the typical benthic ecological approach has been to concentrate on physicochemical characters of the environment and usually to tolerance responses of organisms to these characters through the changing seasons.

Saccoglossus habitat

While much attention has been paid to the study on the marine and estuarine intertidal benthic ecology of temperate zones during the last three decades, our knowledge of tropical coasts is relatively poor. This is particularly true for Indian coasts. Mcintyre (1968) dealt with meiofauna and macrofauna of some tropical beaches of India near the Marine Biological Station, Porto Novo, south of Madras: Ganapati and Lakshmana Rao (1962). Ganapati and Chandrasekhare Rao (1962) dealt with the ecology of intertidal sands of Visakhapatnam coasts Nagabhusanam and Chandrasekhara Rao (1969) and Nagabhusanam (1972) made some preliminary observations on the shore fauna of Orissa coast. Prof. A. Choudhury and the scientists of S.D. Marine Biological Research Institute, Sagar Island, initiated study on benethic fauna of the deltaic Sundarbans since 1965.

Sundarbans, the most sensitive and vibrating mangrove ecosystem is being supported by the numerous delta of the river Ganges the lower stretch of which is known as river Hooghly (vide physiography in Chapter1). Mud flats are the most extensive intertidal habitats in the majority of southern part of the deltaic Sundarbans. In unravelling the complicated network of interactions that exist in the biotic community of these specialised habitats it seems logical to examine it in detail. Where there is little biotic interference, the tidal mud flats are characterised into two major and conspicuous geomorphological units (i) mangrove swamp in the raised part of the delta which is flanked by (ii) the marginal mud flats. These mud flats may be very extensive with very gradual slopes only being covered with a thin sheet of algal mat (blue-green algae) or simply by ever-shifting fluid detritus material. Very sparsely distributed Porteresia grass or few saplings of ***Avicennia*** spp. or ***Acanthus ilicifolius*** are occasionally found on these apparently barren mud flats.

Prentice island is such a type of virgin delta in the Sundarbans complex from where the first specimen of the new ***Saccoglossus*** species has been discovered and collected. Since then this island. among many others harbouring ***Saccoglossus sundarbanensis*** sp. nov. has been chosen for intensive study on the ecobiology of the hemichordate worm. This small island of about six square kilometers area and strewed with several meandering tidal creeks, supports a very attractive ever green canopy of mixed mangroves flanked by sumptuous mudflats the most favourite habitat of the hemichordate worm described in the present thesis (Plate 3, A and B and Fig. 2).

Mangrove vegetation associated with the saltmarsh grass Porterasia coarctata at its margin actually provides with the rich organic detritus food for host of benthic animals occupying the mud floors. Inside the island the forest floor is covered with shedded mangrove leaves which undergo microbial degradation to be reduced to essential nutrients (nitrite, nitrate, phosphate, silicate, free amino acids etc.) and the nondegradable parts contribute to the organic detritus, rich in particulate organic matters. Both the nutrients and the particulate organic matter are mixed up with the substratum sediment formed of clay, silt and partly sand particles. During the receding tides following the hightides, twice in every twenty four hours, nutrients and POM are transported as forest output from the forest floor to the adjacent mudflats and the aquatic subsystem. During this process, which is a daily phenomenon in the dynamic ecosystem, a substantial portion of the nutrients and POM are adsorbed in the fine sand, silt or clay particles, and a part of them is leached down in the interstitial spaces of the substratum. The excess of the nutrients flows out into the coastal water. This is the secret why these apparently barren mudflats of the estuary form the most favourable habitats that support such a diversity of benthic organisms, both phyto- and zoo- and macro, meioand micro. They are either suspension feeders or detritus feeders.

Prentice island has been under continuous surveillance in connection with a multiyear multidisciplinary integrated research programme and hence all necessary information related to the edafic factors (ecological parameters) of the mud flat habitats have been documented through seasons. Texture of the mudflat substratum of different depths contributes to different characteristic habitat niches for particular type of infauna and epifauna (crawlers, creepers or burrowers).

The most common epifauna of the Prentice island mudflats are the Metaplax and ocypod crabs (periodically infauna when retreat in burrows). **Cerithidea** and **Telescopium**, gastropods, barnacles & oysters, mudskippers (**Boleopthalmus**) etc. and the common infauna are the mud mussels (**Anadara, Macoma**), polychaetes. ophiuroides, nemertines, echiurids, mud eels, anemones and hemichordate worms. All these animals depends for their food directly or indirectly, on the mud substratum for their substances. Many of them work and rework in the mud substratum for their living and leaves characteristic trails or markings which help to locate their habitat burrows.

The hemichordate worm ***Saccoglossus sundarbanensis*** sp. nov. is spotted on the mud substratum of the mid littoral zone of the island by its very characteristic faecal castings (Plate 26, b). Just after the ebb tide, as the mud floor slowly gets exposed and evaporates, the hemichordate worms, which feed during flow tide on suspended particulate organic matters, retreat in their burrows. Their tail end is projected through an exit burrow on the surface of the soft substratum from which the faecal matters are deposited in a characteristic coil (Plate 26, Fig. b).

The substratum it prefers as its habitat and the dried feacal castings have been analysed and examined to study the respective textures containing sand, silt and clay as the three major components. The percentage composition of substratum texture throughout the year is as follows: Sand particles varies from 5.4 to 12.4%, silt 51.9 to 56.01% and clay 36.51 to 39.33% (Table 5). The percentage composition of the feacal casting texture (after drying) is as follows: Sand 37.87%, silt 53.98% and clay 8.15%. Other essential ecological parameters of the habitat have been enumerated in the following section.

HYDROLOGICAL PARAMETERS

During the study period from August 1982 to July 1984 seasonal variations of the hydrological parameters, that regulate the Macro and micro climatic conditions of the hemichordate habitat, have been methodically documented (Tables 1 and 2. Figs. 6 and 7).

Temperature

The temperature of atmospheric air, flood water, substratum and interstitial water recorded during the investigation period showed a more or less parallel trend of change through the seasons. The maximum and minimum temperature recorded for air, flood water. substratum and interstitial water were 31.5°C and 24.0°C, 31.5°C and 23.5°C, 33.5°C and 20.0°C. and 31.5°C and 21.2°C respectively.

The range of temperature variations for air were 24.0°C to 31.5°C during 1982 to 1983 and 24.1°C to 31.2°C during 1983 to 1984. 23.5°C to 31.5°C during 1982 to 1983 and 24.0°C to 31.3°C during 1983 to 1984 for flood water; 20.0°C to 33.5°C during 1982 to 1983 and 20.5°C to 30.0°C during 1983 to 1984 for the substratum; and 21.5°C to 31.5°C during 1982 to 1983 and 21.2°C to 30.5°C during 1983 to 1984 for interstitial water.

The temperature was recorded high in the pre monsoon and low during post monsoon seasons and hence an annual bimodel temperature oscillation seemed to be a characteristic phenomenon in this ecosystem. Two maxima, one in the pre monsoon and other in the monsoon months were recorded (Figs. 6, 7 and 8).

Salinity

It is usually a known fact that salinity in the open marine environment does not show any conspicuous fluctuation. The factors affecting salinity are minimum compared to its vast water masses. But in the estuarine environment the salinity is affected tremendously by the additional freshwater freshets from the upstream river and surface run off during monsoon every year. The maximum and minimum salinity recorded for flood water and interstitial waters were 27.9% and 16.87% 28.32%, and 19.92%, respectively during 1982 to 1984. The maximum values was recorded in the months of March and April whereas the minimum value was found during July and August (Figs. 7 and 8).

The pattern of salinity regime through seasons around Prentice island environment is as follows: Maximum in the pre monsoon months (March to June) which accounts no rain but higher temperature, and was minimum in the monsoon months (July to October) affected by South West monsoon winds accompanied by heavy rains. The post monsoon period (November to February) may be considered a mixed period because the water is somewhat diluted with just over monsoonal rains, followed by the winter months with minimum records of temperature and precipitation. During this period, the salinity gradient kept a gradual rising trend and attained a maximum value through the pre monsoon months. Hence, every year the salinity gradient increased to a maximum during the pre monsoon months paralleling the higher temperature and decreased during the monsoon months and was inversely proportional to the rate of precipitation (Figs. 7 and 8).

The interstitial water showed higher salinity values than in the adjacent flood water of the estuary during the pre monsoon and post monsoon periods because of the evaporation of the mudflats during dry seasons. During monsoon, the values were low due to bleaching of excess fresh water.

Dissolved oxygen

The concentration of dissolved oxygen in the flood water and interstitial water showed an oscillation range 2.4 ml/1 to 4.5 ml/l and 0.9 ml/1 to 2.9 ml/1 throughout the study period. It was recorded high in the late pre monsoon and early monsoon and low during early pre-monsoon period.

pH

pH maintained almost a uniformity for both the flood water and interstitial water. The maximum and minimum values for flood water and interstitial water were 8.3 to 8.2 and 7.9 to 7.2 respectively throughout the study period (Figs. 7 and 8).

Population fluctuation of *Saccoglossus sundarbanensis* sp. nov. in relation to variable physical parameters

Monthly sampling of the ***Saccoglossus*** species was made from the intertidal mudflat of Prentice island in Sundarbans which presents a virgin mangrove environment (Table 3). All essential hydrological parameters, which are supposed to maintain and regulate the hemichordate population through seasons, were also documented month wise during 1982 to 1983 and 1983 to 1984 (Tables 1, 2 and 4). Seasonal variation in soil texture and parameters and available carbon were also estimated to correlate with the seasonal abundance of the Saccoglossid population (Table 5). Month wise population study revealed that maximum number per square meter was recorded in the month of November i.e. in the post monsoon season when the average salinity of flood and interstitial water of the environment was about 21% and 24%. Details hydrological parameters of Prentice island mudflathave already been presented in Page Nos. 69 and 71 of the present dissertation and an attempt has been made to draw a correlation between the seasonal population fluctuation of the hemichordate worms and the different parameter variables.

Correlation coefficient between population of *Saccoglossus sundarbanensis* sp. nov. and different ecological parameters

Salinity

Positive relationship (0.202) between ***Saccoglossus sundarbanensis*** and salinity of interstitial water was observed but this relationship was not significant (Table 6A).

Dissolve oxygen

Significant but negative relationship (-0.499) between ***Saccoglossus*** sp. and D.O. was obtained. Number of ***Saccoglossus*** sp. was inversely proportional to dissolved oxygen of interstitial water (Table 6A).

pH

Relationship (0.018) between ***Saccoglossus*** sp. and pH was neither positive nor negative i.e. It was insignificant (Table 6A).

Interstitial water temperature

Population and water temperature show negative but insignificant result (-0.363) (Table 6A).

Hole temperature

Saccoglossus sp. population and hole temperature show significant but negative result (-0.404). It was significant at 5% level (Table 6A).

Anova

Variation of ***Saccoglossus sundarbanensis*** population was insignificant over years (Table 6B).

BEHAVIORAL STUDY

Burrowing and movement

In a soft marine environment all softbodied animals burrow by means of an essentially similar mechanism (Clark, 1964). This is based on the two types of adaptive formations of the body called 'anchors' first by dialation of part of the body above the distal extremity to form what has been termed the 'penetration anchor' and second, by distension of the extremity to produce 'terminal anchor (Trueman, 1968b). These anchorage are applied alternately, first, firm anchorage either of the terminal or penetration anchor, and secondly a movement of one part of the animal further in the substratum.

Soft bodied burrowing animals are generally circular in crosssection or compressed a little dorsolaterally, a shape idially suited for burrowing since the body wall is in contact on all sides and all muscles may contribute to the locomotory force. In an animal of this shape the existence of a true fluid skeleton enables powerful movements to be made since the force developed in muscles remote: from the point of application is transferred to the lower extremity by means of the hydraulic system of the body cavity. This latter feature may be the coelenteron in an anemon, the haemocoel as in pelycepod molluscs, the coelom as in Arenicola (polychaete) and hemichordate worm. These body cavities all function in a similar manner as hydraulic organs (the fluid within acting as 'hydrostatic skeleton') and exhibit their maximal development in respect of the burrowing habit.

Saccoglossus sundarbanensis located in various islands of Sundarbans mangrove complex, inhabits and prefers always soft muddy substratum rich in silt and clay, rich in detritus as well which is a continuous contribution of the mangrove ecosystem. During several unearthing events it was revealed that the animals live in deep burrows, irregularly U-shaped, particularly in case of the larger animals. In a freshly unearthed specimen for its deep burrow and left over the mud substratum just after the receeding tide the long proboscis was found to work and rework with its narrow tip to enter into the soft soil.

During this burrowing process waves of peristalsis were noticed to pass over the proboscis, from the distal proboscis end to the base, disappearing at the collar. Burrowing is entirely accomplished by the proboscis which is elongated first and thrust forward, then contracts by its longitudinal musculature, forming a peristaltic buldge that travels posteriorly at a speed of about 1 mm per 2 or 3 seconds and with a frequency of about 5 or 6 per minute. This observation has been made under stress and it is presumed that the nature and frequency of peristalsis may vary individually at its different state of activity and in natural burrowing process. Meanwhile, before the bulge has reached the proboscis base, the tip is thrust onward again and another bulge follows. These bulges serve to anchor the proboscis while the tip moves onward the body is drawn passively along.

It has been further observed in ***Saccoglossus sundarbanensis*** that it burrows by alternate elongations and contractions of both proboscis and collar, anchoring in the mud by a peristaltic bulge while the body is drawn onward. It has also been observed that this species covers itself with fine silt and clay particles by ciliary action. Cilia drive the silt grains back over the proboscis to thecollar where a mud girdle forms, and this is gradually pushed back by concomitant additions from infornt until the whole animal is covered with mud.

Locomotion

(a) Peristaltic locomotion

In all enteropneust worms, generally described as sluggish and inactive animals, the most active part is the proboscis, constantly engaged in exploring the surroundings. The tip of the proboscis, in case of the present ***Saccoglossus*** species was thrust out, becoming thinner in the process, then it swelled and was withdrawn slightly.

The contraction force generated by the circular muscles which caused the withdrawal travelled posteriorly as a peristaltic bulge, of which both the anterior and posterior edges were equally distinct. Subsequently, the tip was thrust out and withdrawn again, initiating the next wave of contraction. The buldges moved at a speed of about 1 mm per second and with a frequency of about 4 to 5 per minute and the buldges never travel beyond the base of the proboscis. In an actively burrowing specimen, when one bulge reached the base of the proboscis, the next was half way and a third was starting at the tip. When in the mud the buldges must tend to remain anchord while the proboscis moves forward.

Retreat peristalsis

When ***Saccoglossus sundarbanensis*** was kept for observation in a glass aquarium filled with mud taken from its habitat, the proboscis was found to be elongated posteriorly and swelled anteriorly. The swollen part probably was used as anchor, while the elongation of the base helped to push the collar region backwards. A series of waves of contraction were noticed which started at the posterior end of the trunk and travelled gradually anteriorly throughout the trunk, collar and proboscis. The distal anchorage of the proboscis relaxed and became thin as the first wave approached. This was followed by the drawing back of the proboscis.

(b) Movement by ciliary action

When unearthed ***Saccoglossus specimens*** were left on the mud substratum of Sundarbans, their preferred habitats, they were found to move over the mud surface by ciliary action. The posterior end was dragged along and the body became rather elongated. The cilia are primarily responsible for movement within the burrow. Cilia of the proboscis beat steadily posteriorly and never to the reverse direction (observed under microscope keeping small pieces in watch glass filled with sea water).

Saccoglossus sundarbanensis moved over mud surface with the ventral surface downwards, so the strong cilia of the ventrolateral ridges were in contact with the substratum. These are the main locomotor cilia of the trunk. When the trunk was cut in pieces, those small pieces also moved about, showing the direction in which those cilia were beating. The synchronous reversals of the ventrolateral cilia and their co-ordination with peristalsis strongly suggest that they are under nervous control.

Feeding and digestion

Saccoglossus sundarbanensis lives in soft mud of the mangrove floors with an abundance of particulate organic-rich materials. Like all other enteropneusts the present ***Saccoglossus*** species also resorts to suspension feeding, sorting out selected particulate clay materials Impregneted and adsorbed with organic food substances. Though the feeding mechanism as described by various authors (Barrington, 1941; Burdon&Jones, 1952: Knight&Jones, 1953) are more or less same in all the species so far described, personal observation of the present author for the present ***Saccoglossus*** species in a new environment is being communicated in the present dissertation.

Examination of exact feeding mechanism directly in the habitatburrows is very difficult to conduct. Worms after collection were kept in the glass aquaria or taken in petridishes and examined meticulously under zoom binocular microscope. The operation of the proboscis which is the prime body part of the animal and responsible for manipulation of particulate organic rich detritus in suspension through regulated ciliary action.

Feeding is primarily of the mucousciliary type. Detritus and plankton and even the diatoms of the algal mats covering the muddy habitat substratum that come in contact with the surface of the proboscis are trapped in mucous and carried posteriorly by strong ciliary currents. The ***Saccoglossus*** species that inhibit burrows project the proboscis from the mouth of the burrow, moving it around to explore the nutrient rich substratum.

Sand grains or clay particles that touch the proboscis become entangled in mucous secreted by the mucous glands of the proboscis epithelium and the mucous strands are passed back along the proboscis to its base and the collarette. The cilia on the elevated ridges of the preoral ciliary organ at the ventral side of the proboscis stalk and infront of the collar and mouth, beat strongly so as to direct the food particles entangled in mucous strands into the grooves of this organ (Fig. 4, A) and then eventually pass into the mouth. The ciliary organ probably has a chemoreceptive function (Hyman, 1959).

The collarette may sometimes reject some of the mucous strands when the animal does not prefer feeding. The cilia of the proboscis beat steadily backward and never reverse. In the pre-oral ciliary organ, cilia beat from the dorsal ends of the U toward its midventral part, where the formed current joins the general posteriorly directed stream. Fine particles collected in the groove and travelled down the dorsal cornue as thin mucous, strings, which drawn out ventrally to join the main stream of mucous used in food collection. The epithelium of the pre-oral ciliary organ is equipped with numerous cells which have the form of sensory cells and lie above a special tract of nerve fibres (Knight&Jones, 1952). Besides setting up a current into the mouth, the ciliary organ probably functions as an organ of taste.

In this process of ciliary mode of feeding, the gill apparatus also plays a significant role, for its cilia help to maintain a powerful cillary current entering the mouth and existing by the gill pores; however, the gill apparatus does not help to capture food particles.

Freshly collected specimens from the mangrove mudflats show mud loaded gut. As it has been mentioned earlier that the burrow sites are detected by its fresh and characteristic faecal castings comprised mostly of mud and fine sand particles. To extract necessary amount of desired nutrient the animal has to ingest large amount of nutrient rich particulate matters in suspension and entrapped in mucous strings. The digestive tract is a straight epithelial tube that in general lacks intrinsic musculature.

The large mouth is situated ventrally in the collar between the collarette and the proboscis stalk. The mouth opens by the contraction of the radial muscle fibres which traverse the collar coelom at its anterior lip and is closed by an oral sphincter. Mouth leads into the buccal tube which is restricted in the collar and is lined by tall vacuolated, ciliated epithelium often richly furnished with the goblet type of gland cells and having a nervous layer at it base.

The entire gastro-intestinal tract of the ***Saccoglossus*** species is ciliated with differential cell thickenings at different regions of the tract viz., pharynx, oesophagus, midgut and hindgut which have been described in details in Part 1 of the dissertation. A small oesophagus leads into the long straight intestine. Its anterior region is demarcated as the hepatic zone because of the presence of a specialised comparatively much taller columnar glandular epithelial zone at the ventro-lateral aspect of the gut lumen. Posterior to the hepatic region from which it is not definitely demarcated, the intestine continues as a tube of low, often with sinuous ciliated epithelium with a wide lumen.

Digestion

The mucous cords pass as such to the gastrointestinal tract of the trunk via the branchial sector only to be exposed to various digestive enzymes. All three major hydrolyzing enzymes of the digestive system have been quantitatively studied to establish the possible relation between the digestive enzymes and food habit of the animal. The animal shows a strong to moderate amylolytic activity in oesophagus and the second part of the intestine. A very low proteolytic activity, both in acid and alkaline pH, was observed in the second part of the intestine. Esterase was detected in all the zones of the gastrointestinal tract (Table 8).

During the ejection of faecal matters the end part of the trunk segment is pushed outside the burrow and castings are laid down on the mud surface during the ebb forming a characteristic design on the mudflat (Plate 26, B).

Excretion

Excretion in the strictest sense refers to the removal of the waste products of metabolism, which comprises CO_2 and H_20 that are released by the oxidation of energy rich compounds, and the nitrogenous waste that results from the metabolism of proteins and nucleic acids. But the removal of CO_2 is part of the respiratory process, and because of this excretion has commonly been thought of as the removal of nitrogenous waste in particular.

In hemichordate worms this very important biological manifestation has been attributed to an organ called glomerulus, named by Spengel (1983) on the ground of an apparent resemblance to the glomeruli of vertebrate kidneys. The glomerulus was generally accepted as an excretory organ in hemichordate from the time of Bateson (1885). Koehler (1886) and Scheimkewitsch (1888), although experimental proof in favour of its excretory functioning was lacking. Macro and micro-anatomical details of the glomerular apparatus, as is found in ***Saccoglossus sundarbanensis*** have already been communicated in Chapter I, page 25 to 34. (Plate 9).

In the present dissertation work, the author was unable to demonstrate experimentally the real excretory function of the glomerulus but tempts to surmise from the histological details and cellular architecture of the glomerular tubular evaginations associated with proliferated pigmented cells customarily called nephrocytes or chloragogen cells. That the glomerulus in the ***Saccoglossus*** species shares the responsibility of removal of nitrogenous wastes from the body is evident from its intimate relation with the blood sinus and the haemal system. The author is not very certain from his observation how and where the waste products are removed. This important physiological problem demands further intensive study.

Respiration and Osmoregulation

Respiration is a complex physico-chemical phenomenon by which exchange of$O_2 \rightleftarrows CO_2$, takes place between the organisms/organic tissues and the surrounding environment. This is external respiration. Secondly, there is the complex reactions that take place within the cells with the ultimate evolution of energy by oxidation of the energy rich molecules derived from food; this is internal respiration. Each living organism terrestrial, fresh water and marine in habit, must accomplish this physiological manifestation by unique devices of their own.

In the present ***Saccoglossus*** specimen, as portrayed in the previous Chapter, the pharynx of the animal is differentiated into a dorsal branchial region and a ventral digestive region, the later being reduced to a mid ventral strip (Hypobranchial ridge). The branchial region of the trunk segment of the acorn worm is perforated on each side by longitudinal series of openings, the gill-slits, 85 to 126 pairs on an average, oriented dorsoventrally. The pharyngeal slits are curved in conformity with the curvature of the lateral pharyngeal wall (Plate 19).

The gill-slits in the adult ***Saccoglossus*** , when completely developed and transformed with the downward growth and advancement of the secondary or tongue lamellae, the gills appear to open through a U-shaped aperture. In the acorn worms the gill slits do not open directly to the exterior instead, into the pouch-like cavities, the branchial sacs, which in turn, open on the body surface by the gill pores located in the longitudinal dorsolateral branchio-genital grooves.

It has also been projected in the earlier Chapter that the vascular system of ***S. sundarbanensis***, like all other enteropneusts, isa open vascular system. The vascular plexus of the pharyngeal wall supplies branchial apparatus through fine lacunar spaces. During the respiratory operation, the important functional characteristics of the present ***Saccoglossus*** species is the production of the respiratory current by the combined actions of muscles and cilia abounding the pharyngeal slits and the gill pores. Oxygenated water is imported into the buccal cavity through oral aperture and transported into pharynx through ciliary action, and from the pharynx to the exterior via the branchial sacs and gill pores. Thus the whole operation is so synchronised that the maximum respiratory efficiency is maintained within the limits of the Hemichordate plan of structure. Joint physiological role shared by the respiratory, circulatory and locomotory systems of this enteropneust worm provides an Interesting example of organisational integration in favour of this fragile animal that each system is complementary to the successful functioning of the other.

Osmoregulation

Data obtained for the body fluids of a wide range of organisms show that the great majority of marine invertebrates have an internal osmotic pressure that agrees very closely with that of the sea water in which they live (Barrington, 1967). This is one reason why it is believed that life originated in the sea, or at least became associated with it at an early stage. It is assumed that protoplasm might have been at one time carrying on its metabolic activities while being directly bathed by sea water.

Osmotic relationships, however, are not the only ones that have to be considered in this context, the ionic composition of the body fluid is no less important. It is now well recognised that marine invertebrates, despite being usually iso-osmotic with the sea, maintain an ionic composition of their body fluids markedly different fromthat of the normal sea water. ***Saccoglossus sundarbanensis*** while living in its burrow and being bathed with imported sea water during each high tide, maintains by diffusion more or less some chlorinity in its body fluid as per with the outside flood water. Another important phenomenon is the direct sea water import in the proboscis coelom through the proboscis pore.

Breeding Behaviour and Biology

Saccoglossus sundarbanensis inhabit the fluvial mudflat just above the lowwater mark spring tides to a little way above low water mark neap tides. During the breeding season between February and June, when salinity of the environmental water reaches between 20 to 28%, ripe male and female worms are kept in large glass troughs with little amount of mud and sufficient brackish water from the habitat. After a week of acclimatization in artificial environment a gentle rubbing with needle on the gonadal ridges of the ripe worms initiates them to spawn.

In ripe male and female worms gonads become conspicuous in colouration and trunk morphology. In both the sexes testes and ovaries are lodged in paired lateral gonadal ridges and are arranged in paired peudo-somatic saccular configurations. The gonads do not have any gonoduct but open in the branchio-genital grooves by their gonopores. The gravid females when they spawn in the laboratory glass trough, at least 200 to 300 eggs are released from each sacculated ovary. Ripe males, kept in petridishes with sea water from the habitat. when irritated, ejaculates white milt in the immediate water environment. Following Colwin and Colwin (1953) released eggs were collected in watch glasses filled with filtered sea water collected from the habitat and left for 4 to 6 hours in small groups of about 100 in separate batches. Released sperms collected from male specimens were washed repeatedly with filtered sea water. Then the batchesof eggs in separate watch glasses were inseminated artificially with the sperms so obtained.

Microscopical observations revealed that large number of sperms adhered on the periphery of the vitelline membrane of the mature eggs (Plates 25, C & D). It is only in those eggs which were devoid of germinal vesicle, after the egg-sperm physical contact and interaction, the cortical membrane was found to be elevated, a phenomenon known as activation of the egg. This is an evidence of successful fertilization of eggs as it stimulates and initiates the first meiotic nuclear division. First polar body extrusion has been observed under the microscope. Unfortunately, synchronous early cleavage stages could not be documented in the laboratory, may be due to some obvious disadvantages in maintaining the desired environmental parameters in the experimental watch glasses. The author could not get hatched tripartiate juveniles in the laboratory as it was achieved by Colwin and Colwin (1951) and Burdon&Jones (1953). The present author intends to conduct intensive study on this developmental part of ***S. sundarbanensis***after this dissertation.

As development is direct in most saccoglossid worms and the Just hatched tripartite larvae pass a very small time in planktonic phase and settle in the substratum to crawl and make a fresh burrow to grow within, the author has to collect the post hatched juvenile stages from the hemichordate habitats of the mangrove environment with the help of plankton nets (bottings silk no. 25). Plenty of saccoglossid juveniles of different developmental stages have been collected from natural habitats and examined. In most of the cases the completely ciliated larvae exhibit three distinct divisions with conspicuous proboscis segment and a telotorch cilialy band but no apical tuft of cilia.

Saccoglossus sundarbanensis sp. nov. and its associated macrobenthic fauna in Sundarbans mangrove mud flats

This dissertation intends to portray the benthic community, both epifauna and infauna, those coinhabit the enteropneust worm ***S. sundarbanensis***in the Prentice island mudflat.

Studies on these communities have primarily been concerned with understanding the distribution of population and/or communities in space and have shown the importance of physical factors (e.g. sediment mobility with seasonal erosion or acretion) in controlling distribution patterns (Dexter, 1979). Tentative quantitative analysis of seasonal changes in the structure of the physically stressed communities have also been carried out and communicated (Table 8).

Physical characteristics of the sample sites which represent the habitats of the hemichordate worms and associated macrobenthic fauna of Prentice Island mudflat have been studied in detail and communicated in this section of the dissertation. Elevation above MLW and sediment characteristics (Fig. 5, Table 5), which control spatial distributional patterns of macrofaunal communities inhabiting this mudflat, did not change significantly with season.

The mean annual salinity in the flood water adjacent to the sample sites was always lesser than 30%. During periods of high monsoonal precipitation the flood water salinities often dropped below 20% (Table 1, Fig. 6). Such low salinity regime was of short duration, generally lasting only for 50 to 60 days and confined to low phases of the tidal cycle. Salinity of intertitial water did not generally differ much from the flood water salinity.

Temperature (sediment and water) varied in the expected fashion with season. Lowest flood water temperatures of approximately 23°C were recorded during 1982 to 1983 and highest flood water temperatures of approximately 31.5°C were recorded during 1982 to 1983. Sediment (burrow/hole) temperatures followed a similar seasonal cycle with a low of approximately 33°C during 19821983. In addition, sediment temperature varied with depth below the surface.

Composition of Benthic Community

Thirty four species of macrobenthic invertebrates and one species of vertebrate benthic fauna (Boleopthalmus) were obtained during the survey. Table 7 summarises the associated faunal survey data including the hemichordate.

Among these associates, very few are suspension feeders, e.g. two members of actiniarians (***Pelocoetes exul and Edwarsia zonesi***).

others are either detritivores or deposit feeders viz., the bivalves (***Macoma birmanica, Anadara granosa***), a holothuroid (unidentified) a polycheate (***Capitellethus disper***). The rest are grazers feeding on leaf litters or benthic algal mat. These are several species of brachyuran crabs listed in Table 7 and few species of gastropods viz., ***Cerithidea cingulata, Telescopium telescopium***. The Nemarteans and many polychaetes are foraging predators. The community is a unique assemblage of animal species covering different invertebrate and vertebrate plyla of which ***Saccoglossus sundarbanensis*** sp. nov. is a conspicuous infaunal component.

With all, the other associate members of the community the hemichordate worm, a suspension feeder, also shows its marked population fluctuation in response to the changes in seasonal parameters (Fig. 9), the maximum density obtained during the months of November and December. Post monsoon is the season when saccoglossid juveniles are recruited naturally in numbers in the Prentice island mudflats. mostly in the midlittoral belt.

Table 1. Hydrology of the flood water August 1982 to July 1984

Month	Temperature water °C	pH	D.O. (ml/1)	Salinity (
August	30.3	8.2	3.6	16.87
September	30.3	8.2	3.39	17.2
October	30.0	8.3	3.8	19.35
November	28.0	8.2	2.8	21.3
December	28.9	8.3	2.86	22.38
January	23.5	8.2	3.3	24.75
February	28.4	8.2	3.4	24.21
March	30.0	8.3	2.4	26.36
April	30.5	8.2	2.5	27.9
May	31.3	8.2	4.4	23.8
June	31.5	8.2	4.5	23.2
July	31.5	8.2	4.5	22.8
August	29.5	8.2	3.4	17.58
September	31.0	8.3	3.9	18.48
October	29.3	8.2	3.6	19.3
November	27.5	8.2	3.2	20.10
December	25.7	8.2	3.2	21.9
January	24.0	8.3	3.2	23.68
February	27.0	8.2	3.0	24.78
March	28.5	8.2	2.4	26.00
April	30.9	8.2	2.5	26.7
May	31.2	8.2	3.2	26.32
June	31.3	8.2	3.8	24.6
July	31.3	8.3	4.5	23.8

Table 2 : Hydrology of the Interstitial water August 1982 to July 1984

Month	Temperature		pH	D.O. (ml/l)	Salinity (%.)
	Burrow (°C)	Water (°C)			
August	30.5	30.0	7.4	1.32	24.39
September	30.9	30.5	7.2	1.32	24.9
October	30.0	29.5	7.5	2.9	25.25
November	22.5	25.0	7.5	0.98	26.32
December	21.0	24.0	7.5	0.9	26.7
January	24.0	25.5	7.8	0.9	23.31
February	20.0	21.5	7.4	0.98	23.8
March	33.5	31.5	7.8	0.89	28.32
April	27.5	26.0	7.9	2.3	27.25
May	24.5	23.9	7.8	2.2	24.8
June	25.9	23.5	7.5	2.2	23.3
July	24.5	24.5	7.4	1.9	22.4
August	28.5	28.5	7.5	1.70	19.92
September	29.9	30.0	7.4	1.30	20.99
October	25.9	27.5	7.5	1.2	21.8
November	22.0	25.0	7.5	0.98	22.61
December	21.2	24.0	7.5	0.97	23.33
January	20.5	21.2	7.8	0.98	24.04
February	23.5	23.5	7.8	0.89	24.39
March	28.0	29.0	7.2	0.90	26.89
April	30.5	30.5	7.4	0.98	26.18
May	25.0	24.9	7.9	1.90	24.04
June	24.5	24.5	7.4	2.20	22.25
July	25.0	24.8	7.4	2.30	19.92

Table 3: Population of hemichordate worm per sq. meter

August 1982 to July 1984

Month	Lower mid litoral zone	Middle mid litoral zone	Upper mid litoral zone	General average population
August	1.10	4.13	1.83	2.35
September	0.55	4.16	0.38	1.69
October	1.00	3.00	2.00	2.00
November	3.66	11.0	4.30	6.32
December	2.11	8.00	3.31	4.66
January	0.66	5.00	2.33	2.99
February	2.93	4.16	2.63	3.26
March	2.66	8.38	2.38	4.47
April	2.30	3.50	2.00	2.60
May	1.60	3.30	1.90	2.36
June	1.33	2.66	1.00	1.64
July	1.30	3.66	1.66	2.31
August	1.33	4.33	1.66	2.66
September	0.33	4.0	0.66	1.66
October	2.0	7.5	2.48	3.99
November	3.00	10.8	3.52	5.77
December	2.16	8.0	3.32	4.49
January	0.66	5.0	2.33	2.66
February	0.66	5.33	2.33	2.74
March	2.66	4.33	2.66	3.21
April	1.66	5.33	2.0	2.99
May	1.33	3.66	2.0	2.33
June	2.0	5.55	2.55	3.35
July	0.66	6.33	2.0	2.99

Table 4 : Air temperature and rainfall (average per month)

Month	1982-1983		1983-1984	
	Air Temp. (°C)	Rainfall (mm)	Air Temp. (°C)	Rainfall (mm)
August	30.00	20.00	29.00	18.68
September	29.00	2.77	28.50	12.57
October	30.20	6.47	27.50	8.93
November	27.00	1.07	26.50	0.34
December	30.00	-	25.30	0.27
January	24.00	0.87	24.10	0.38
February	27.50	1.31	26.50	0.14
March	28.00	2.07	27.00	-
April	31.00	0.03	30.50	0.08
May	30.50	2.24	30.50	1.60
June	31.50	11.17	31.20	15.20
July	30.50	5.08	30.00	9.20

Table 5: Seasonal variation of various parameters in Prentice Soil (1982 – 1983)

Soil	Sand (%)	Silt (%)	Clay (%)	pH	E.C. (m.mhos/cm)	Organic carbon (%)
December 1982 to February 1983	5.42	55.19	39.33	7.3	5.35	0.64
March 1983 to May 1983	6.22	56.01	37.77	7.3	5.48	0.69
June 1983 to August 1983	12.40	51.09	36.51	8.2	3.24	0.68
September 1983 to November 1983	9.82	52.77	37.41	8.3	3.29	0.64

Table 6: Correlation coefficient ® between Saccoglossus sp. pulation density (no. / m^2) and different ecological parameters

	Water temp. (°C)	Burrow temp. (°C)	pH	D.O. (ml/1)	Salinity (‰)
Saccoglossus sp.	-0.363	-0.404*	0.018	-0.499**	0.202

B] Anova for Saccoglossus sp. population.

Source of variation	Degree of freedom	Sum of square	Mean of square	'F' value
Month	11	33,50	3.05	7.44**
Year	1	0.31	0.31	0.76
Error	11	4.48	0.41	-
Total	23	38.29		

*Significant at 5% level.

**Significant at 1% level.

Table 7: Number of associated fauna per square metre area in the mud flats of Prentice Island through seasons during 1982 – 1983 and 1983- 1984

Name of species	Pre-monsoon	Monsoon	Post-monsoon	Pre-monsoon	Monsoon	Post-monsoon
Pelocoetes exul	10.25	9.5	14.5	13.5	13	24.5
Edwardsia zonesi	1.75	2.25	3.75	2	3.25	4
Nemertinea species	4	2	2.5	4.75	2	3.75
Lumbrineris polydesma	9	5	6	16.5	7.5	11
Lumbrineris impatiens	3	13.5	9	4.5	15	8.5
Lumbrineris heteropoda	-	1.5	3.5	0.5	4	6.5
Heteromastus similis	8	3.5	3	7.5	2	2.5
Capitellethus disper	1.5	11.5	6	4.5	10	6.5
Scyphoproctus djiboutensis	-	1.5	0.5	-	2	-
Nereis chingrighattensis	2.5	7	5.5	4	9.5	7.5
Dendronereis heteropoda	4.5	2	3	7	1.5	4
Namalycastis fauveli	-	2	4	0.5	1.5	3.5
Glycera tesseleta	10.5	3.5	9	10.25	6.5	13
Glycera lancadivae	-	4	7.5	0.5	2	5
Diopatra cuprea cuprea	3	1.5	3	2.5	3.5	0.5
Talehsapia annandalei	-	0.25	2	0.5	1	0.5
Phyllodoce gracilis	1	-	2.5	2.5	1.5	4
Cirrolana sp.	3	16.25	9.5	6	12.5	7
Cloridiopsis immaculata	-	0.5	0.25	1	0.75	2
Mud shrimp	-	1.75	0.5	-	-	0.5
Ilyoplax gangetica	12	16.5	28	14.5	19.5	26
Dotilla blanfordi	16.5	8.5	20	15.5	13	22.5
Macrophthalmus sp.	0.5	1	1.5	-	-	2
Metaplax intermedia	20	16	7	13	5.5	6.5
Dotillopsis brevitarsis	11.5	4.5	7	13	5.5	6.5
Uca dussumieri dussumieri	2.5	0.5	1	1.5	0.5	0.5
Neosolen aquedalcioris	3.5	1.5	0.5	2	1	1.5
Macoma (Psmmacoma) birmanica	3.0	3.25	4.0	2.5	3	3.5
Anadara (Anadara) granosa	5	4.5	5.5	4.2	4	4.5
Assimnia brevicula	7.5	12.5	8.5	5	14	8.5
Cerithidea (Cerithideopsis) cingulata	2.5	6	4.5	1.5	5.5	5
Telescopium (Telescopium) telescopium	0.5	1	0.75	-	1.5	0.5
Boleopthalmus bodderti	-	0.5	1	0.5	1	0.5
Holothuroid	10.5	10.25	16	12.5	9.5	18.25

Table 8. Major hydrolying enzymes of *Saccoglossus sundarbanensis* sp. nov.

	Amylase mg/maltose 10 mg protein	Esterase mg/naphthol 10 mg protein	Alkaline protein Tyrosin equivalent/10 mg protein	Acid protein Tyrosin equivalent /10 mg protein	Invertase mg glucose/10 mg protein	Cellulase mg glucose 10 mg protein
Branchial zone	1.165	0.116	0.031	0.11	0	0
Oesophageal zone	2.66	0.20	0.071	0.122	0	0
Intestine I (= Hepatic zone)	2.03	0.305	0.083	0.1457	0.028	0
Intestine II (Post hepatic zone)	4.7	0.295	0.1815	0.15	0.0125	0

Table 9. Fatty acid composition (% W/W) of *Saccoglossus* sp. nov.

Component acids[a,b]	Percentage (W/W)
12 : 0	0.3
14 : 0	1.7
14 : 1	0.3
15 : 0	0.3
16 : 0	19.7
16 : 1	14.1
17 : 0	2.8
17 : 1	0.9
18 : 0	6.1
18 : 1 W9	5.0
18 : 2 W6	0.6
18 : 3 W3	1.1
20 : 3 W9	2.0
22 : 0	0.2
20 : 4 W6	13.2
20 : 2 W9	10.2
20 : 5 W3	11.3
22 : 4 W3	0.9
22 : 5 W6	1.4
22 : 5 W3	2.4
22 : 6 W3	2.7

[a]First and second figures represent carbon chain length : number of double bonds.

[b]The co-values represent the carbon chain lengths between methyl group and nearest double bond to it.

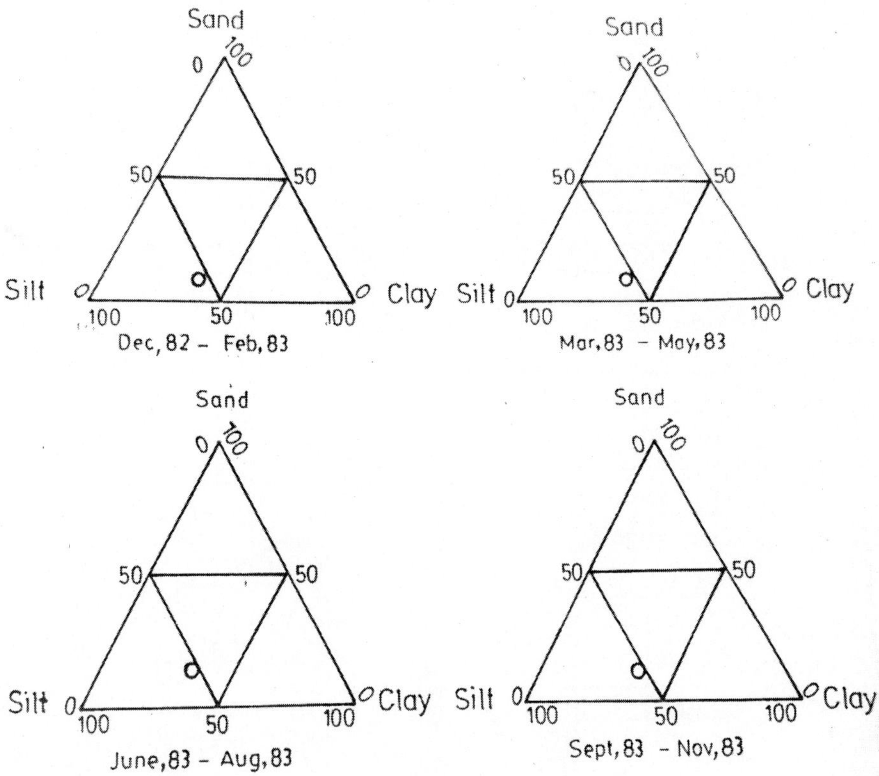

. SOIL TEXTURE OF THE MUDFLAT IN PRENTICE ISLAND

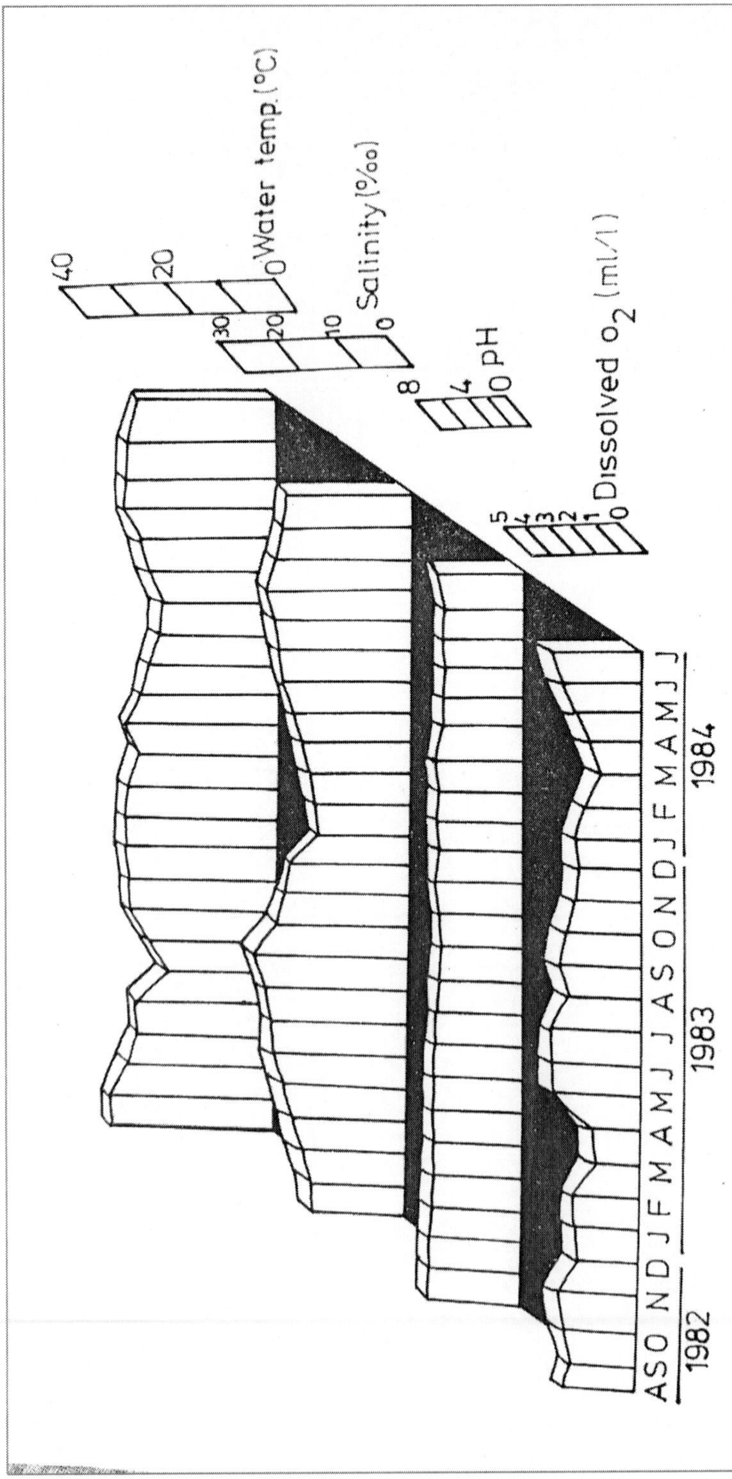

Fig.6. HYDROLOGICAL PARAMETERS OF FLOOD WATER OFF PRENTICE ISLAND

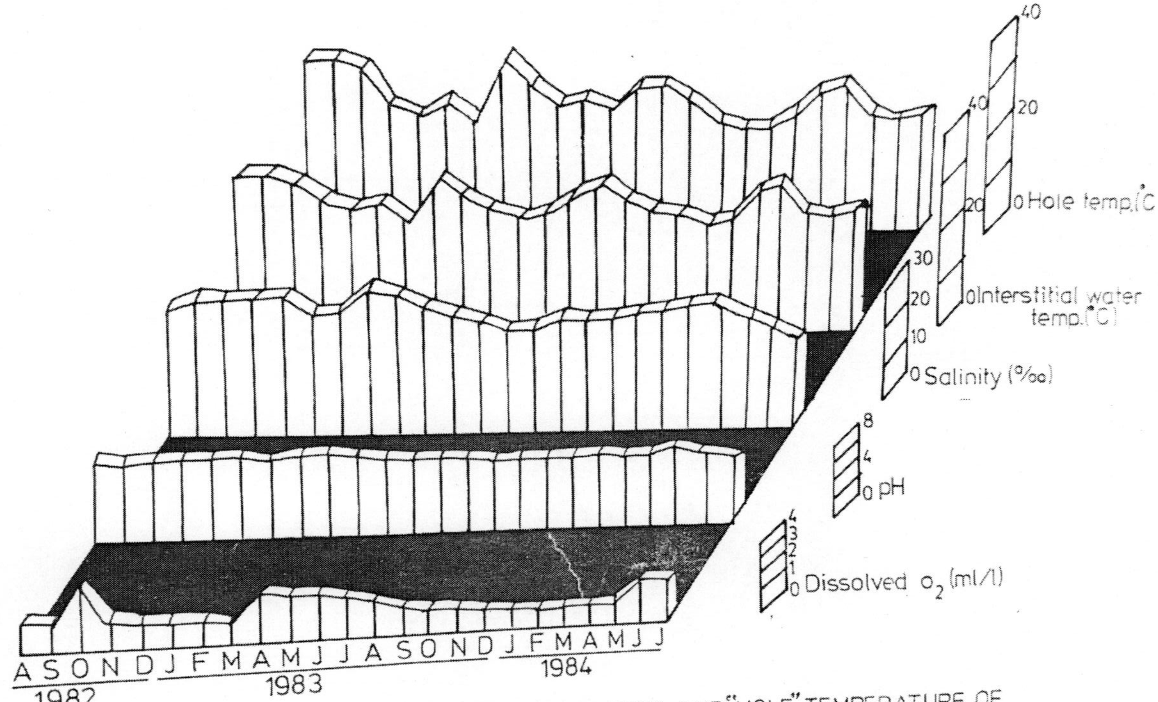

HYDROLOGICAL PARAMETERS OF INTERSTITIAL WATER AND "HOLE" TEMPERATURE OF THE MUDFLAT IN PRENTICE ISLAND.

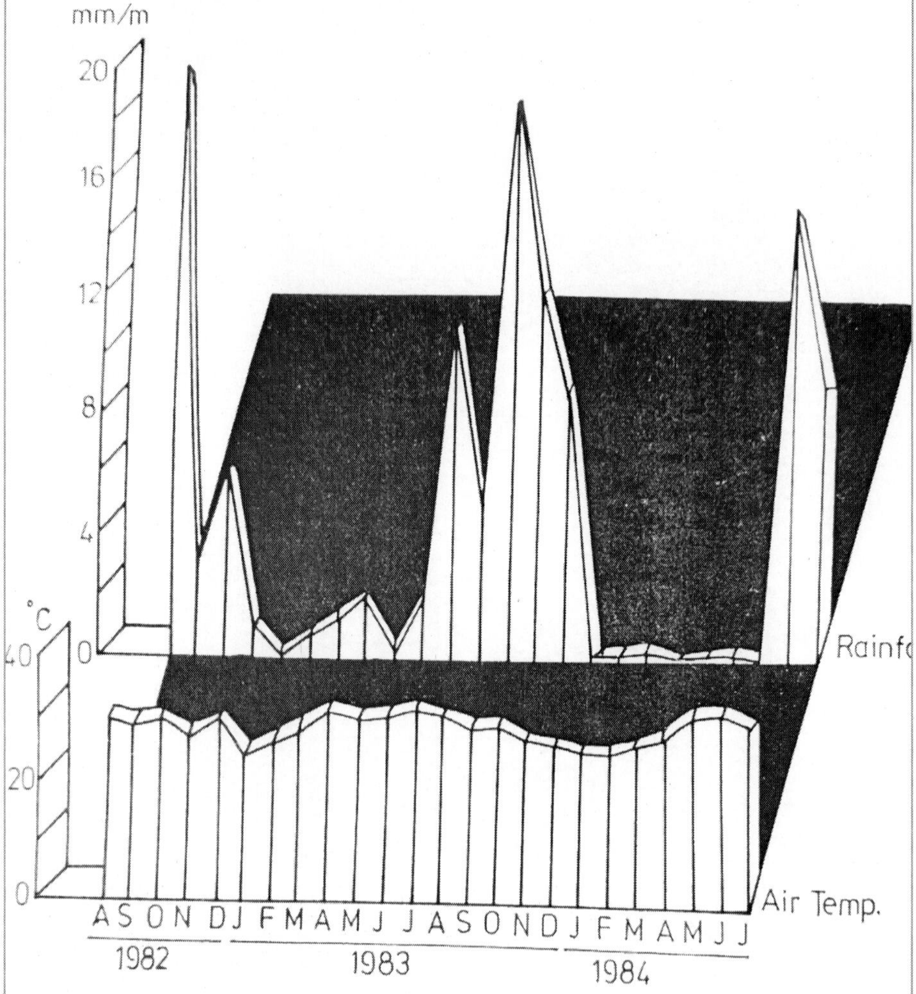

Fig.8. ATMOSPHERIC TEMPERATURE AND PRECIPITATION RECORDS IN WESTERN SECTOR OF SUNDARBAS.

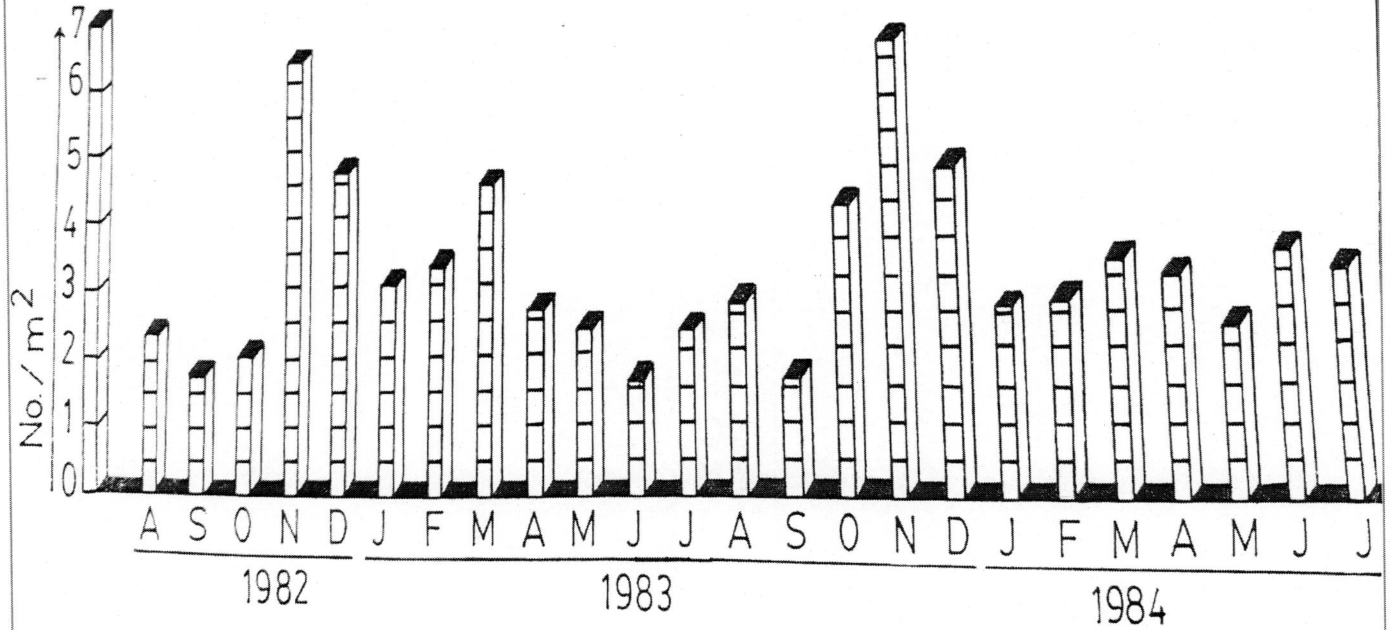

Fig.9. Population density of *Saccoglossus sundarbanensis* sp. nov. m^{-2}

ANNEXURE

Lipid synthesis in *Saccoglossus sundarbanensis* sp. nov. as energy source to maintain chemoecological relationship

In course of the ecological study of the enteropneust hemichordate worm in this saltmarsh mangrove environment it was felt to examine its fatty acid composition and detailed lipid chemistry which may highlight the problems of the multitudes of estuarine benthic animals and plants exposed to continuous environmental stresses of the vibrating mangrove ecosystem. Organic fat is confined to the animal and plant triacylglycerols which are supposed to be the major dietary components and important storage of energy as well in the animal community.

RESULTS AND DISCUSSION

When the pooled lipid extracts from ***Saccoglossus*** sp. were diluted with large volume of water, a heavy white precipitate appeared at the interface of the aqueous and nonaqueous solvents. This observation was similar to that made by Ghosh ***et al.*** (1985) which occurred during the extraction of leaf lipids of mangroves.

On examination by thin layer chromatography (TLC), using various standards and spray reagents, the white substance responded to the colour reactions of terpenes and sugars. From this observation It may be suggested that, the substance is a glycoside in probability. Accumulation of terpene glycoside in this species, makes it an interesting and important subject matter for future study.

Fatty acids profile (Table 9) of ***Saccoglossus*** sp., as reported by Singh ***et al.*** (1987), indicated that altogether 22 components were present, of which total unsaturated acids (68%) were much higher than total saturated acids (32%). Of the polyunsaturated fatty acids, level of arachidonic acid (20: 4 W6) was highest (13.2%), which was in contrast with the previous works (Misra ***et al.***, 1983; Dutta ***et al.***. 1986) on other detritivorous benthic animals of this ecosystem, whereas eicosapentaenoic acid (20: 5 W3) was found in all the animals, as the major component polyunsaturated fatty acid.

Another biologically important fatty acid, viz.. eicosatrienoic acid (20 3 w9), which is found in all the benthic animals of this ecosystem, was also found in this animal in low levels, as usual. Most interesting observation was the presence of docosadienoic acid (22: 2) in appreciably high levels (10.2%), compared to 1.2% in other benthic animals sharing the same ecosystem. Presence of two double bonds in this acid was confirmed by hydrogenation data, but regarding their positions, the acid should be isolated and should be confirmed.

From the present investigation on the fatty acids, of ***Saccoglossus*** sp., it is apparent that, this species exhibits some characteristic features, which is absent in other benthic animals of Sundarbans and makes the species an important biological materials for further biochemical studies.

PLATE 24

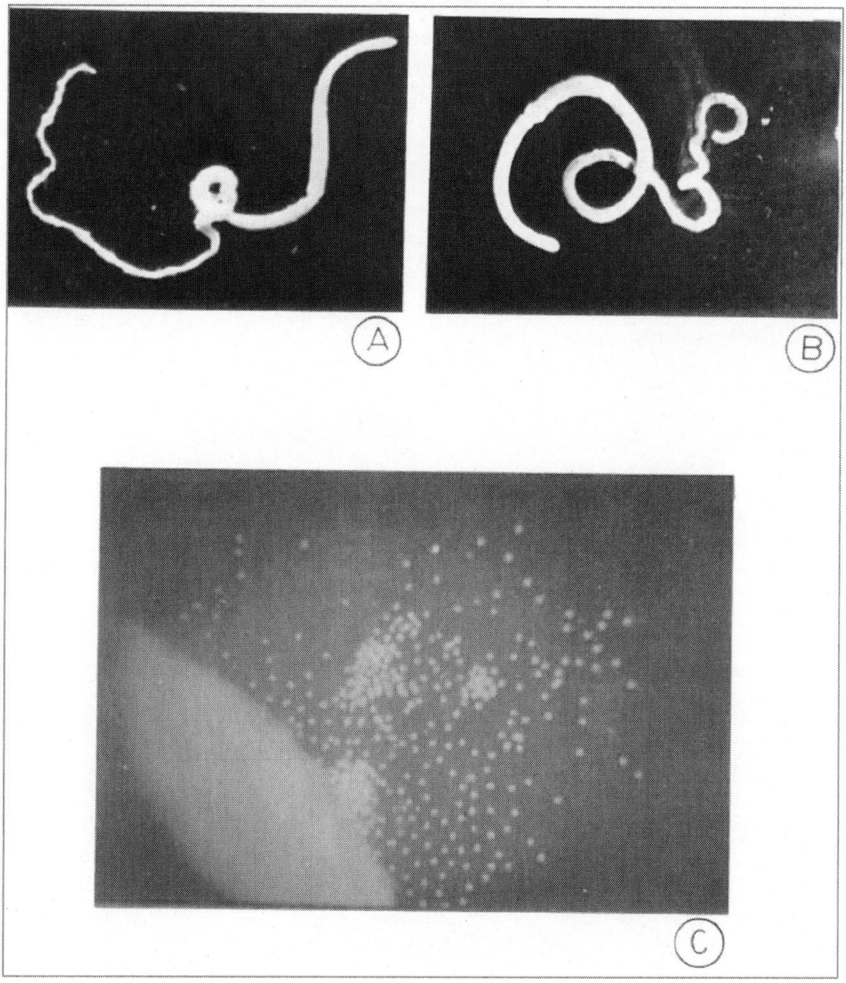

(A) Female ***Saccoglossus sundarbanensis*** sp. nov.

(B) Male ***Saccoglossus sundarbanensis*** sp. nov.

(C) Female ***Saccoglossus*** worm releasing eggs.

PLATE 25

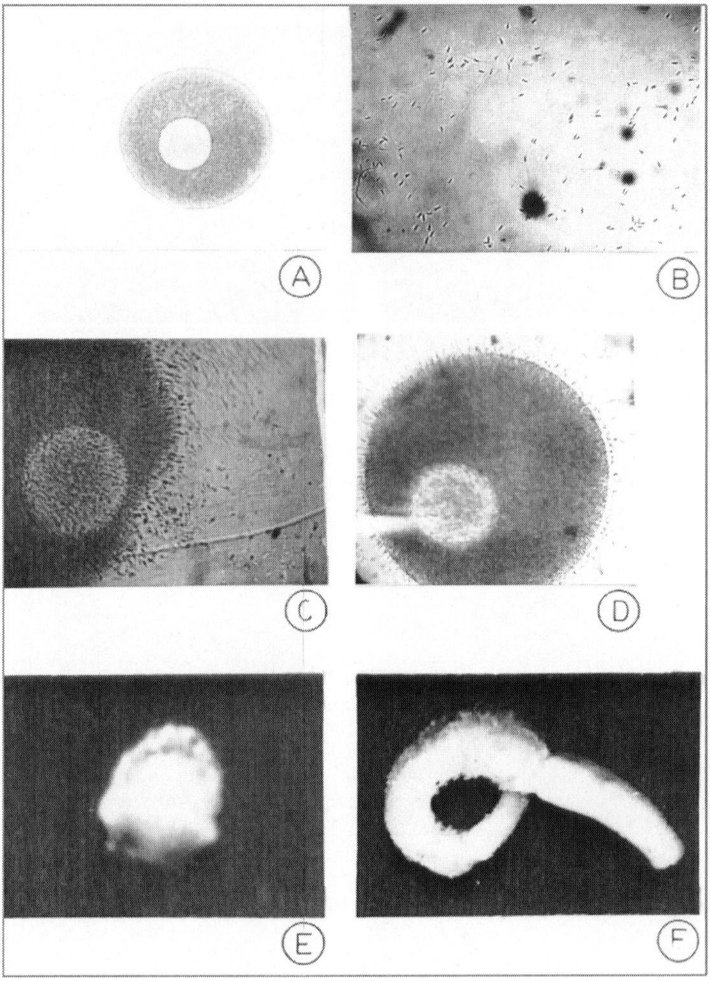

(A)Mature egg ready for fertilization in laboratory.[10 X 10 μm]

(B) Released sperms with long tails. [10 X 100 X 1.25 μm]

(C) Egg and sperms mixed together. [10 X 100 μm]

(D) Event of fertilization. Sperm egg interaction. [10 X 100 μm]

(E) Tripartite larva collected from flood water. [15 X 20 μm]

(F) Juvenile ***Saccoglossus*** worm collected from the mud flat of Prentice Island. [15 X 20 μm]

PLATE 26

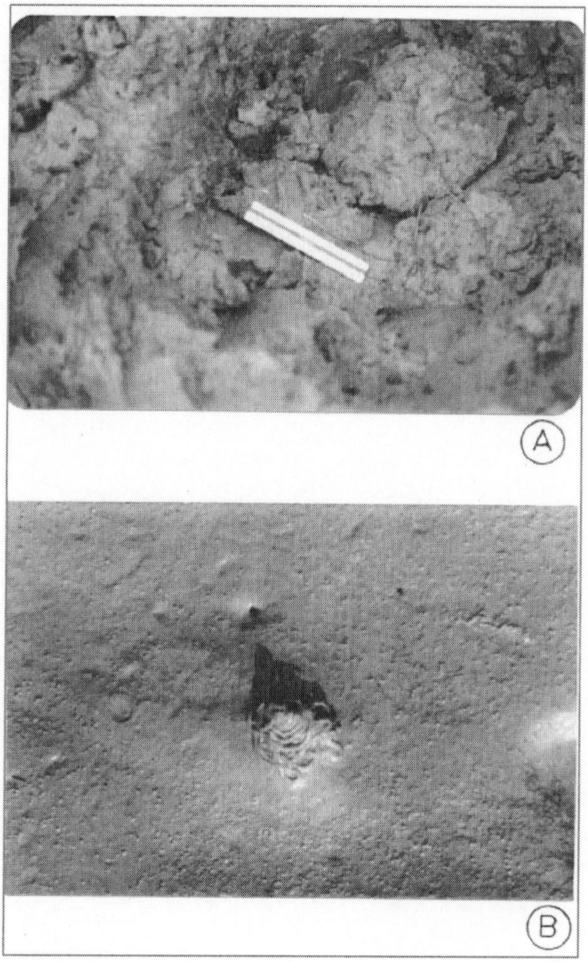

(A) Freshly unearthed intact ***Saccoglossus*** worm. It measures nearly 76 cm in length.

(B) Casting of ***Saccoglossus*** worm on the mud flat and a pore beside through which proboscis is protruded.

PLATE 27

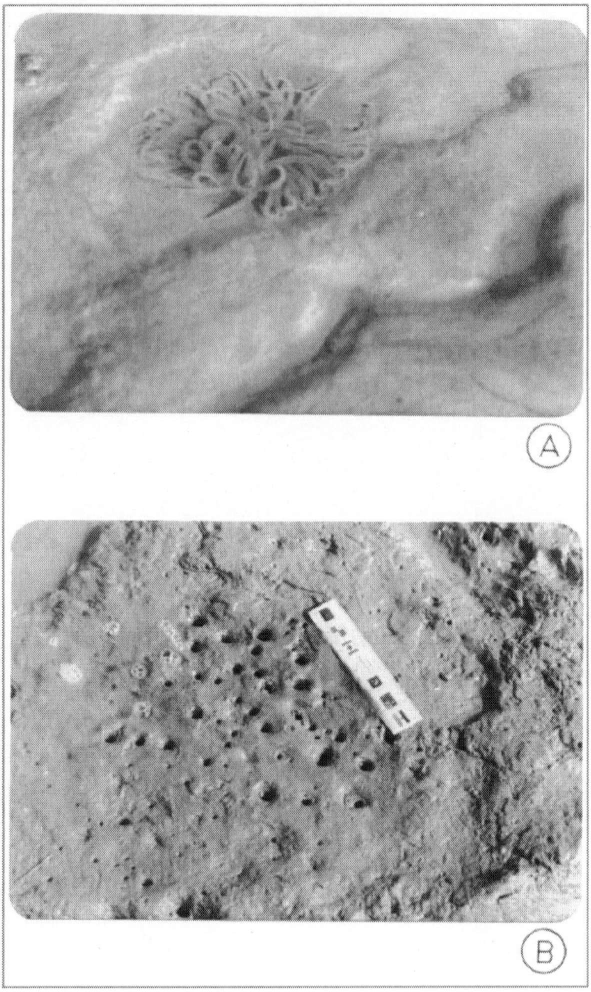

(A) Head crown of the sea anemone ***Paracondylactis ramunii***.

(B) Burrow openings of ***Pelocoetes exul***.

PLATE 28

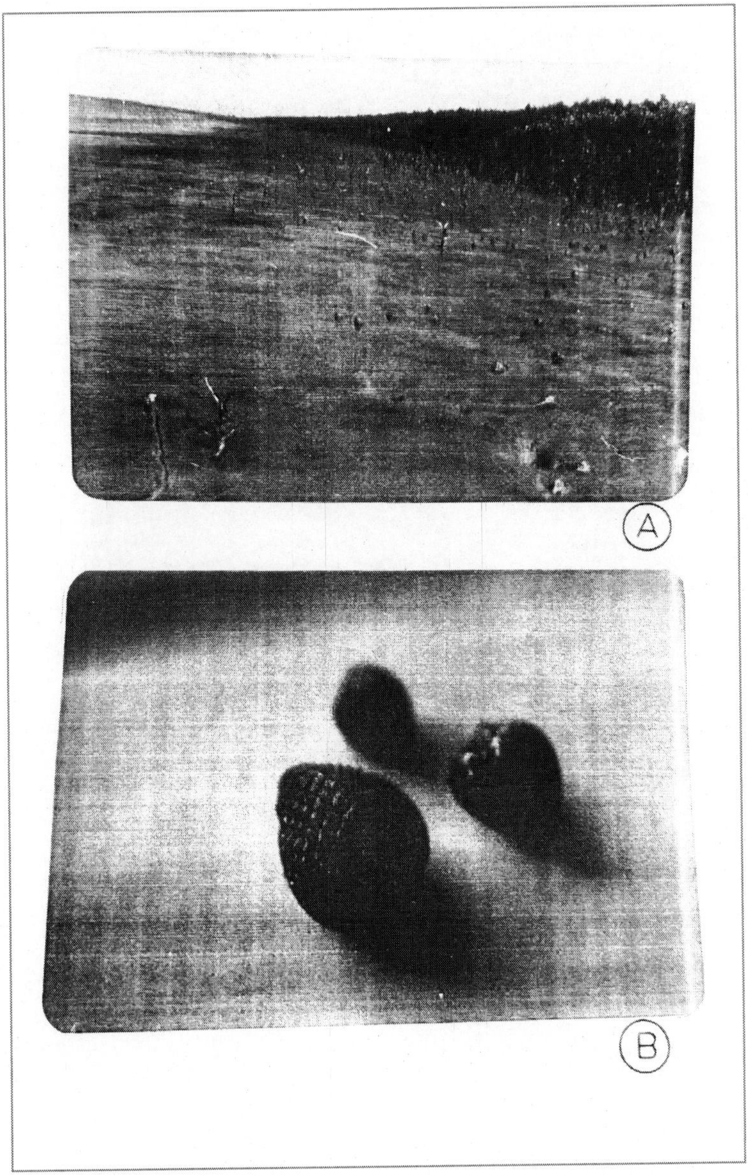

(A) ***Telescopium telescopium*** feeding on algal mat on the exposed mud flat during low tide.

(B) ***Anadara granosa*** embedded in mud during low tide.

PLATE 29

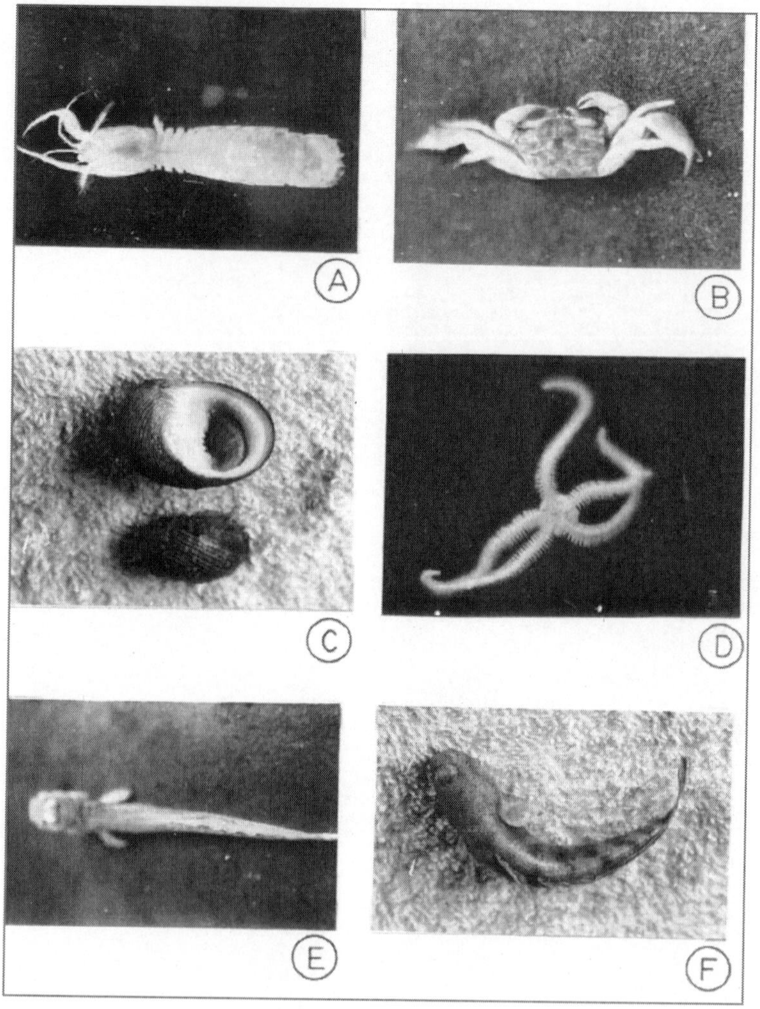

(A) Cloridopsis bengalensis
(B) Pea crab
(C) Nerita articulata
(D) Brittle star (unidentified)
(E) Boleopthalmus viridis
(F) Gobiopsis macrostomus

PLATE 30

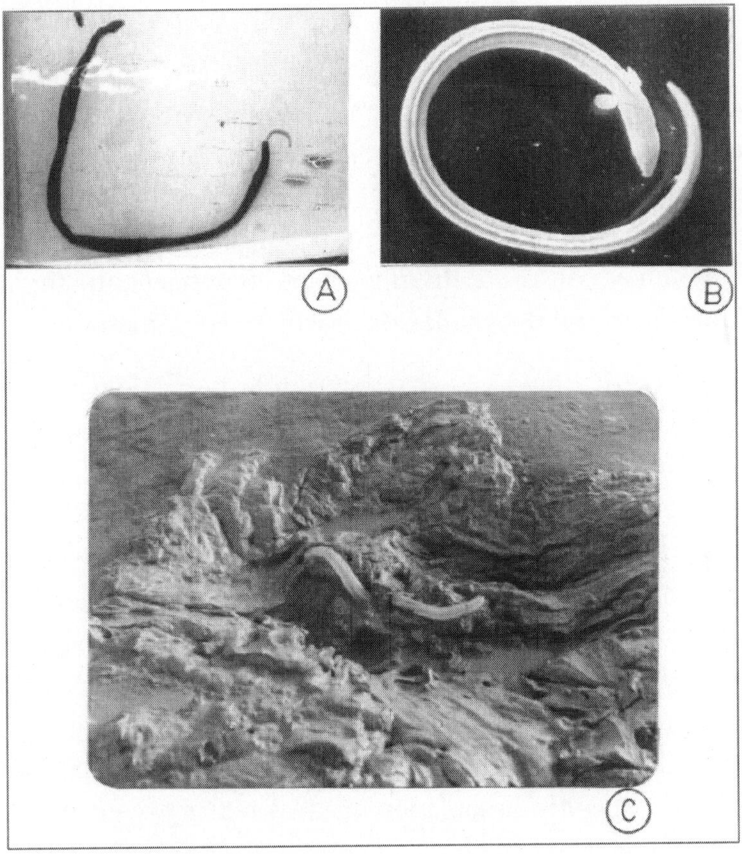

(A) Nenartine worm with extruded proboscis.

(B) Pisodonophis sp.(mud eel)

(C) An apodid holothuroid (unidentified)

DISCUSSION

Enteropneust hemichordate worms are natural macrobenthic fauna of the marine and estuarine sediments, some are common in the intertidal mudflats. The concept of intertidal benthic community in the tropical estuarine environment are better realised during the recent times as benthos constitute the important components of the aquatic ecosystems at several trophic levels in which they function.

Intertidal mangrove mudflats are very rich in faunistic wealth but are recognised as the most rigorous environments of the tropics. Not only the organisms inhabiting these areas are exposed to wide range of physical stresses (e.g.. desiccation, temperature variation. salinity, unusual weather phenomena, sediment transport, tidal currents etc.). but they also experience a full range of biological interactions (e.g.. competition and predation). Critical research analysis on the quantitative structure of the macrobenthos provides substantial information to realise various materiological and hydrological impact on the abundance of benthic population, on the colonisation of the same or heterospecific association, and also the organism-sediment relationship in the tropical belt.

As herbivores and carnivores, the benthos feed on living algae, plants and invertebrates, while themselves form the main food of many fishes and birds. As detritivores indirectly they help in recycling of organic matter (Brinkhurst and Jamieson, 1971). The benthos are good indicators of water quality because, unlike plankters, they form stable communities in the sediments which integrate changes over long time intervals and which reflect both the sediments and the water column (Cook and Johnson, 1974). Their production, species composition, spatial distribution and temporal variations, and various life history parameters (growth rate, number of generations per year. emergence patterns, etc.) of individual species are important and interesting problems to study. However, problems on benthic ecology. particularly in the HooghlyMatla estuarine complex, have not beenstudied much, and seem to be lagging behind to those on plankters specially with respect to physiologically oriented experiments.

The benthos is a dynamic community (Brinkhurst, 1974) which reacts quickly to environmental fluctuations, specially in a very vibratile ecosystem like deltaic estuarine complex of Sundarbans. It is more likely that an interaction of physical, chemical and biological factors and behavioral mechanisms (Paterson and Fernando, 1971; Segerstrale, 1978) may control the distribution of a species in a particular ecological niche (Sephton ***et al.***, 1980).

One of the very important factor related to the conspicuous characteristic of any estuary is its subjection to periodical exposure and inundation, the timings of which vary in accorddance with the lunar cycle and periodicity. All benthic creatures residing here tune and orient all their biological activities, movement and feeding in particular, with the tidal rhythms.

Prentice island and its mudflat environs are no exceptions to this estuarine climax and may be considered to be a typical model representative of this famous tropical estuary. The island is endowed with a good canopy of mixed mangrove forest in its upper midlittoral ridges and being fringed by the most potential grass belt, Porteracia coarctata, it enjoys to have anall season detritus rich muddy substratum which supports a diversity of benthic fauna and algal mattings. Two years documentation (August 1982 to July 1984) of the meteorological and hydrological parameters (Tables 1, 2 and 4) projects the conspicuous seasonal cycles and ossilations of temperature, precipitation and salinity regimes, though dissolved oxygen and pH showed moderate variations. An array of benthic invertebrates residing in the muddy substratum, micro, meioand macro, and exposed through seasons to all these environmental variables, demonstrate their marvellous osmoregulatory adaptations to withstand the rigorisities of this vibratile ecosystem.

Saccoglossus sundarbanensis sp. nov. a resident of the mudflat of Prentice island and an efficient burrower, reacts to this changing panorama displaying a variety of its behavioural performances. The ***Saccoglossus*** dugout or burrow is not exactly U-tube in this case but is slightly curved and is slightly zigzag instead of horizontal in position, lying somewhat 15 to 40 cm deep in the mud for the adult worms. The burrows are engineered by the combined and coordinated efforts of the proboscis, primarily. The worms make their dugout in preferred soils, silty clay and rich in humous in the mid to lower littoral zone. Inner wall of the burrow is very smooth as the mud and sand particles are glued to the mucous secreted by the epidermal gland cells of the proboscis, collar and trunk.

Though living in the mud house the animal seems to be bit sophisticated and keeps its burrow cleaned by flushing water out by the ciliary action and by depositing its faecal casting outside the burrow forming characteristic and very specific identifying marks (Plate 26B).

The method of burrowing by hemichordate worms has been described by Knight&Jones (1952) for ***S. cambriensis*** and by Rao (1954) for Ptychodara flava both of whom emphasize that the most active part of the body in burrowing is the proboscis which extends by the contraction of circular muscles aided by a dorso-ventral muscle sheet and possibly the diagonal arrangement of the longitudinal muscle fibres (Hyman, 1959).

Examination of the movement in the ***saccoglossid*** worm is a fascinating one which can only be watched precisely in glass aquaria filled with mud or in moderate size glass petri dishes. Major movement is initiated and localised in the proboscis which is the principal exploratory organ of the animal. Initial movement which starts at the tip of the proboscis is passed on to the rear end of the organthrough series of waves of peristalsis. Collar exhibits very little movement activity, but peristalsis does pass through the trunk segments which become eminent when the body is dragged forward.

As the animal is efficient in its forward movement and backward retreat, both forward and retreat peristalsis feature in the locomotory behaviour of ***Saccoglossus sundarbanensis***. Peristaltic locomotory waves are utilised both for movement along a burrow and during the burrowing process in most animals of circular cross-section. Movement by ciliary action is another important criteria in the animal's locomotion, particularly in its juvenile stage.

Heterotrophic benthic organism, marine or estuarine, are either predominantly suspension feeders or deposit feeders, with the exception of a few species that are fecultative feeders, ***Saccoglossus sundarbanensis*** lives in a typical mangrove mudflat on the eastern flank of Prentice island which is common in the deltaic HooghlyMatla estuarine system. This mudflat is shared and co-inhabited by several such benthic invertebrates and vertebrates, both epifauna and infauna, who resort either to suspension feeding or deposit feeding, or simply grazing on the algal mat at the substratumwater interface.

The hemichordate worm under communication is an infauna living in burrows and predominantly is a suspension feeder. Food collection is comparatively very simple and lacks the refinements that are a testimony to the much higher level of functional specialization. Feeding is primarily of the mucousciliary type. During tidal interaction the surface sediment of the detritus rich muddy substratum gets disturbed and silt and clay particles impregnated and adsorbed with particulate organic matters are brought into suspension. These particulate matters along with the diatoms in suspension are caught in the mucous cord generated by the mucociliated epithelium of the proboscis and passed over to the collarette by the ciliary currents. The food laden mucous cord is passed into the buccal opening by the guided ciliary action of the pre-oral ciliary organ.

The ingested food material travel the gastro-intestinal tract which is ciliated throughout, by peristaltic movements caused initially by the circular muscles of the oesophagus. Waves of local contraction travel forwards in this region but the effect is to move food backwards, for at the point of contraction the wall projects inwards as a fold that moves a short distance posteriorly and then returns anteriorly on relaxation. Thus, to follow Knight&Jones analysis. the contained food is impelled backwards by folds of the oesophageal epithelium that acts like the blades of paddled steamers.

The arrangement in the present ***Saccoglossus*** species seems better adapted for the propulsion of solid material, for the posterior portion is moved fast and so makes room for the passage of material further backward. The most interesting feature, in comparison with many other ciliary feeders, is the limited possibility of manoeuvring the food particles once they are within the alimentary canal.

Efficient digestion demands that food and enzymes shall be brought into intimate contact by thorough mixing. A range of digestive enzymes are present which must be elaborated by the regionally differentiated Intestinal epithelium. The limited information available indicates a capacity for digesting starch, glycogen, maltose, casein, sucrose, and triacetin in case of Glossobalanus minutus (Barrington, 1941). This aspect has not been thoroughly investigated by the earlier workers in relation to enteropneust digestive physiology (Barrington, 1965).

In the present investigation the author made an attempt to reveal the presence of various digestive enzymes in the different regions of the intestinal tract to which the food is exposed. Three major hydrolyzing enzymes have been quantitatively measured in the digestive system of the present ***Saccoglossus*** species to establish the possible relation between the digestive enzymes and the food of the animal.

During the present investigation the author felt that more detailed in situ experimental exercises were needed to understand thoroughly the digestive physiology of the enteropneust worms.

So far the part of the dissertation dealing with respiration, osmoregulation and excretion of the present ***Saccoglossus*** species under communication, the author pledges to submit his bumble confession that the entire exercise on these topics has been made based on field observations and histomorphological studies of the related organs instead of physiological experiments. The mechanical operation of branchial trunk segment and the synchronous operations of the individual gill slits have been meticulously examined under the dissecting binocular microscope but the volume of water manipulated in a given time period and the rate of discharge of the used water during each respiratory exercise could not be ascertained.

The elaborate vascular arrangements in the branchial segment and their connection to the dorsal and ventral trunk vessel's help to surmise the possible transport of deoxygenated and oxygenated blood in the circulation. But the author is of firm conviction that apart from the branchial respiration, the hemichordate worm also resorts to cutaneous respiration like the annelids. This is being contended based on the repeated observations conducted on the shivered body parts (viz., proboscis, trunk segment behind the branchial zone, and the caudal segment) without the branchial segment which survived and showed live activities for days when kept in normal seawater. The integument possibly supplemented by diffusion the required oxygen to the shivered body tissues to meet their requisite metabolic demands when they were deprived of the branchial circulation.

To establish an acceptable concept on this very important biological manifestation and to fill up the lacuna in our existing knowledge, the author deems it necessary to conduct detailed experimental investigations which he was unable to perform during the course of this dissertation work. Exactly similar contention is being held by the author whether the responsibility is shared by the glomerulus for the removal of the nitrogenous wastes for the entire animal.

Investigations on the breeding behaviour and reproductive biology of the enteropneust worms was first initiated by Beteson (1884, 1886). He reported first the spawning behaviour and the manner of sperm release in ***Saccoglossus kowalevskli***. Devis' study on ***S. pusillus***

(1908) and Kirk's brief note on ***S. otagoensis*** (1937, 1938) were the only records of early embryology of those enteropneusts with 'direct' form of development (i.e., without tornaria larva). Colwin and Colwin (1953) first conducted an experimental study in the Marine Biological Laboratory. Woods Hole, Massachusetts, on the normal embryology of ***Saccoglossus kowalevskii*** upto the formation of the first stage larva with the first pair of gill pores. Burdon&Jones (1953) added to the detailed observations on the developmental biology of ***Saccoglossus horsti***, particularly of the so called larval stage.

Very few authors, and in very few cases, could make their observations on the breeding behaviour and release of eggs and sperms right in the worm's respective burrows. The present author tried his best but in vain to observe the mature worms to spawn in nature, i.e. in the Sundarbans mangrove mud flat environment. But spawning activities of ripe ***Saccoglossus sundarbanensis*** and release of sperm in the male worms, were studied with satisfaction in the field Laboratory of Susama Devichoudhurani Marine Biological Research Institute. Sagar Island.

Attempts were made in the field laboratory following Colwin and Colwin's method of artificial Insemination after obtaining eggs and sperms from the matured worms. Only partial success was achieved through the elevation of the cortical membrane during egg-sperm physical interaction, initiation of the first meiotic division and extrusion of the polar bodies. The author intends to conduct more detail experimental manoeuvring to elucidate the fertilization events and embryogenesis in ***S. sundarbanensis***in near future.

In the ecological studies on benthic communities, distinction is being made between its structure and function. Structure is defined as fluctuation in space and time of number of individuals and species and function as the type and amounts of energy flowing through the system (Lindeman, 1942).

Benthic community structure of the mud flats of Prentice island has been investigated during the past five years. It has been found that benthic macrofauna of Sundarbans showed conspicuous seasonal variations in their faunistic structure and compositions due to several factors like beach configuration, rate of accretion and erosion, soil constituents and hydrological parameters (salinity, pH, D.O., temperature etc.). The dominant groups of the macrofauna are represented by Hemichordates, polychaeta, cniderians, gastropods, bivalves, nemertines, Ophiuroids, holothuroids decapod crustaceans, goboid fishes and eels.

However, since most tropical estuarine invertebrates, including the enteropneust hemichordates, are reproductively active only seasonally it is possible to hypothesize that the structure of communities numerically dominated by individual organisms would vary seasonally depending upon their recruitment and mortality (due to competition and predation) trends. The object of this study was to demonstrate and evaluate quantitative temporal changes for a muddy intertidal community numerically dominated, sometimes by enteropneust hemichordates at both population and community levels. To facilitate accomplishment of this objective. the author selected a community for which spatial distribution patterns had not previously been documented and described.

It was suggested by Sanders (1956) that two general patterns were typical for benthic communities, Firstly, there is persistance as shown by communities dominated by species like ***Saccoglossus sundarbanensis*** sp. nov. or the bivalve Macoma birmanica that work and rework the sediment and thereby control the range and numbers of coexisting species. Secondly, there are oscillatory patterns shown by Polychaetes, brachyuran crabs, some gastropods and bivalves.

Sanders (1956) also postulated that near one environmental extreme, in the intertidal belt, the fauna is subjected to environmental factors that fluctuate to an unpredicatable manner and, as many species are not able to tolerate these upredictable fluctuations, the species complement becomes low at least at some specified seasons. However, competition and predation effects also operate and these lead to large fluctuations in population sizes and density.

The survey analyses of the Prentice island mudflat community through years suggest and endorse Sanders' contentions that the dominance pattern in different groups of the community change seasonally subjected to the changes in environmental factors. In fact biological interactions of competition and predation may also be very severe and may be playing important role in bringing these seasonal fluctuations.

The important point is that niche specialization in the intertidal belt do not occur since the environment is constantly fluctuating in this ever vibrating estuarine system.

One of the most obvious relationships between the benthic communities and environmental factors is the interacting system between feeding types and the silt-clay fraction of the sediment. Sanders (1958) was the first to quantify this relationship, from his studies of the benthic fauna of Long Island Sound, New York.

The relationship between trophic group distribution and structure of the sediment-water interface is apparent when the faunal data are compared with the sediment profile. All sampling stations in the silt-clay facies are densely populated with deposit feeders specially tellinid or prosobranch bivalves (***Macoma sp.***) and Holothuroids. This environment is also co-inhabited by the hemichordate enteropneust worms that feed on the organic detritus made afloat by the tidal disturbances.

SUMMARY

The dissertation deals with the study on the morphology, ecology and biology of a new hemichordate worm ***Saccoglossus*** ap. (Singh and Choudhury, 1984) which has been discovered from the mangrove littoral mud flats of deltaic Sundarbans, India.

The entire observation has been presented in two Chapters: Chapter1 deals with the morphology containing macroand microanatomy from functional point of view, and ChapterII deals with the biology and ecology of the hemichordate worm.

It has been realised from the macromorphological study of the tripartite animal that the everretractile muscular proboscis of the hemichordate worm is indispensable for this animal in absence of a true head structure. The proboscis plays an important role in the essential biological manifestations of the animal including feeding and burrowing.

The most spectacular revelation of the present investigation is the presence of two protocoels (proboscis coelom) instead of one in the proboscis anatomy. This important character alone, alongwith many other salient features, helped to identify the enteropneust hemichordate worm from Prentice island mud flat to be new to science. It has been compared and verified meticulously with other members of the genus ***Saccoglossus*** of the family Harrimaniidae and has been assigned a new name ***S. sundarbanensis*** .

Macroand microanatomy of the animal specially the proboscis complex ciliary organ, branchial apparatus, digestive tract and the gonadal structures have been dealt with in details.

Related to the biological study of ***S. sundarbanensis***the dissertation portreyed at length the digestive physiology and breeding behaviour of the animal, but the author stressed on further detailed experimental studies on these topics.

In presenting the ecological considerations, identification and physiography of Prentice island, the prime study station, have been protrayed.

The seasonal variation of different environmental parameters viz., air and soil temperature, rainfall, salinity, temperature, dissolve oxygen, pH of flood water and interstitial water, organic carbon, percentage of sand silt clay of the soil substratum in particular, have been documented.

Population fluctuation of ***S. sundarbanensis***in relation to variable environmental parameter has been studied through two successive years and communicated in the dissertation.

Lastly, the macrobenthos forming the community associates of the hemichordate worm, ***S. sundarbanensis***have been documented. The periodical dominance of different members of the community in different seasons in response to changing environmental variables, has been projected.

BIBLIOGRAPHY

Ackman, R.G. and Burgher, R. D. 1965. Cod liver oil fatty acids as secondary reference standards in the GLC of polyunsaturated fatty acid of animal origin : analysis of a dermal oil of the Atlantic leather back turtle. J. Am. Oil Chem. Soc., 42, 38.

*Agassiz, A. 1864. Notes on the embryology of starfishes (Tornaria). Ann. Lyceum Natur. Hist., New York, 8:

Agassiz, A. 1872. Tornaria, the young stage of Balanoglossus. Am. Nat. 6: 636-637.

*Agassiz, A. 1873. The history of Balanoglossus and Tornaria. Mem. Amer. Acad. Arts Sci., 9 (Pt 2): 421-436.

Annandale, N. 1907. Fauna of brackish water ponds of Port Canning, Lower Bengal. Rec. Ind. Mus., 1:

Annandale, N. 1922, Marine element in the fauna of the Ganges. Bijdr. Dierk., 1 : 143-154.

Ansari, Z. A., Harkantra, S.N. and Nair, S.A. 1977. Benthos of the Bay of Bengal: A preliminary account. Mahasagar Bull. natn. Inst. Oceanogr., 10 (1 & 2): 55-60.

Ashworth, R.B. and Cormier, M.J. 1967. Isolation of 2, 6-dibromophenol from the marine hemichordate Balanoglossus bimiensis. Science, 155: 1558-1559.

Assheton, R. 1908. A new species of Dolichoglossus. Zool. Anz. 33: 517-520.

Atamanova, M.V. 1977. Dermal and intestinal epithelium of Enteropneusta as a phylogenetic stage in the development of the epithelia of chordates. Arkh. Anat. Gistol. Embriol.. 73 (9): 55-61.

Atamanova, M.V. 1978. Ultrastructural characteristics of dermal epithelium in Saccoglossus mereschkowskii (Enteropneusta). T. Sitologiya, 20 (12): 1355-1359.

Azariash, J., Ismail, M.M. and Najib, M.A. 1975. Investigation on the ecology and respiratory responses of the hemichordate Ptychodera flava to tidal cycles and salinity changes. Biol. Bull, 149 : 455-466.

Balasubrahmanyan, K. 1959. Occurrence of Glandiceps sp. and Saccoglovesus sp. (Enteropneusta) from Inshore water of Portonovo. Curr. Sci., 28(9) : 371-172.

Barnes, R.S.K. 1977. The Coastline, Chichester: Wiley.

Barrington, E. J. W. 1937, Structure and function of the digestive system in Amphioxus (Branchiostoma) lanceolatus. Phil. Trans. Roy. Soc., B 30: 269-312.

Barrington, E. J. W. 1941. Observation on feeding and digestion in Glossobalanus minutus. Quart. J. Micr. Sci., 82: 227-260.

Barrington, E. J. W. 1965. The biology of Hemichordata and Protochordates. Oliver and Boyd, London.

Barrington, E. J. W. 1979. Hyman's comparative vertebrate anatomy, Wake. M.H. (Ed.). 3rd. University of Chicago Press, Chicago ILL. U.S.A., London, England, Illus.

Barrington, E.J.W. and Thorpe, A. 1963. Comparative observations on iodine binding by Saccoglossus horsti Brambell & Goodhart and by the tunic of Ciona intestinale (L.). Gen. Comp. 3: 166-175.

Bateson, W. 1884. The early stages in the development of Balanoglossus. Quart. J. Micr. Sc., 24: 208-236.

Bateson, W. 1885. The later stages in the development of Balanoglossus kowalevskyi with a suggestion as to the affinities the Enteropneusta. Quart. J. Micr. Sci., 26 : 81-122.

Bateson, W. 1886. Continued account of the later stages in the development of Balanoglossus kowalevskyi and on the morphology of the Enteropneusta. Quart. J. Micr. Sci., 26 : 511-533.

Benito-Jesus, 1982. "Hemichorda" In Synopsis and classification of living organisms. McGraw Hill Book Company, S. P. Parker (Ed.) 2 : 819-821.

Benito-Jesus, 1975. Histological and ultrastructural study of the body wall of Glossobalarius minutus (Kowalewsky) Enteropneusta. Boll. R. Soc. Espanota Hist. Nat. (Biol.), 73: 207-228.

Benito-Jesus, 1975. Histological study of the epidermis of Glossobalanus minutus (Kowalewsky) (Ptychoderidae; Enteropneusta; Hemichordata). Bol. R. Soc. Espanola Hist. Nat. (Biol.), 73: 209-235.

Bhavanarayana, P.U. 1975. Some observation of the benthic faunal distribution in the Kakinada Bay. In recent researches in estuarine biology. R. Natarajan (Ed.). H.P.C. (I) Ltd. NDLH., pp. 146-150.

Bhunia, A.B. and Choudhury, A. 1981. Observation on the hydrology and quantitative studies on benthic fauna in a tidal creek of Sagar Island, Sunderbans, West Bengal. Proc. natn. Acad. Sci., India 47(3): 398-407.

Bjornberg, T.K.S. 1959. On Enteropneusta from Brazil. Bolm Inst. Oceanogr., S. Paulo, 10 : 1-104.

*Blanxland Benham, W. 1899. Balanoglossus otagoensis. Quart. Jour. Micro. Sci. 13:

Boigh. E.G. and Dyer, W.J. 1959. A rapid method of total lipid extraction and purification. Can. J. Biochem. Physiol., 37: 911.

Bose, B.B. 1956. Observations on the hydrology of the Hooghly estuary. Indian J. Fish., 3(1): 101-118.

Brambell, F.W.R. and Cole, H.A. 1934. Occurrence of an Enteropneusta in wales. Nature Lond. 133: 913.

Brambell, F.W.R. and Cole, H.A. 1939a. The pre-oral ciliary organ of the Enteropneusta: Its occurrence, structure and possible phylogenetic significance. Proc. Zool. Soc., London., B 109 : 181-193.

Brambell, F.W.R. and Cole, H.A. 1939b. Saccoglossus combrensis sp. n. an Enteropneusta occurring in wales. Proc. Zool. Soc. Lond., B 109 : 211-236.

Brambell, F.W.R. and Goodhart, C.B. 1941. Saccoglossus horsti sp.n. an Enteropneust occurring in the solent. Journ. Mar. Biol. Ass., U.K., 25: 283-305.

Brinkhurst, R.O. 1974. The benthos of lakes. MacMillan, London and Basingstoke.

Brinkhurst, R.O. and Jamieson, B.G.M. 1971. Aquatic Oligochaeta of the world. Oliver and Boyd, Edinburgh, p. 860.

Bullock, T.H. 1940. The functional organization of the nervous system of Enteropneusta. Biol. Bull., 79 : 91-113.

*Bullock, T.H. 1944. The Giant nerve fibre system in Balanoglossus. J. Comp. Neurol., 80 : 355.

Bullock, T. H. 1945. The anatomical organization of the nervous system of Enteropneusta. Quart. J. Micro. Sci., 86 : 55-111.

Burdon-Jones, C. 1950a. An Enteropneusta genus new to the British Isles. Nature, 165 : 327.

Burdon-Jones, C. 1950b. Records of British Enteropneusta. Nature 165 : 636-637.

Burdon-Jones, C. 1951. Observation on the spawning behaviour of Saccoglossus horsti Branmbell and Goodhart and of other Enteropneusta. J. Mar. Biol. Ass., U.K., 21 : 625-638.

Burdon-Jones. C. 1952. Development and Biology of the larva of Saccoglossus horsti (Enteropneusta). Phil. Trans. Roy. Soc. London, B 236: 553-559.

Burdon-Jones, C. 1956. Observation on the Enteropneusta Protoglossus koehleri (Caullery and Mesnil). Proc. Zool. Soc., London, 127: 35-58.

*Burdon-Jones, C. 1956. Nachtrag : Enteropneusta. In Handbuch der zoologie (Kukenthal and Krumbach, Eds.). 3rd 2 (9): 57-78. de Gruyter, Berlin.

*Burdon-Jones, C. 1957. Hemichordata, Family Ptychoderidae: Tornaria larvae. Fiches d' identification du zooplankton. Sheet 70: Cons. Int. Explor. Mer.

Burdon-Jones, C. 1962, The feeding mechanism of Balanoglossus gigas. Bid. Foc. Fil. Cien. Letr. Univ. S. Paulo, No. 261. Zoologia 24: 255-280.

Burdon-Jones, C. and McIntyre, A.D. 1960. Stereobalanus a Genus new to the old world. Nature Lond. 186 : 491-492.

Burdon-Jones, C. and Patil, A.N. 1960. A revision of the genus Saccoglossus (Enteropneusta) in British water. Proc. Zool. Soc. Lond., 134 : 635-645.

*Burne, G.C. 1889. On a tornaria found in British seas. J. Mar. Biol. Ass., I: 63-68.

*Caulley, M. and Mesnil, F. 1916. Sur un Enteropneuste (Dolichoglossus kowalevskii). Bull. Soc. Zool. Fr. 41 : 125-127.

Carrikar, M.R. 1967. Ecology of estuarine benthic invertebrates: A. perspective. In "Estuaries". Lauff, G.H. (Ed.). American Association for the Advancement of Science, Washington.

Chakraborty, S. K., Choudhury, A. and Deb, M. (1986). On Decapod Brachyura from Sundarbans mangrove estuarine complex, India. Journal of Bengal Natural History Society: New series, 5 (1): 55-68.

Choudhury, A., Das, A., Bhattacharya, A. and Bhunia, A.B. 1984a. A quantitative assessment of benthic macrofauna in the intertidal mudflats of Sagar Island, Sundarbans. Proc. Asian Symp. on Mangrove Env. Res. and Management, Kaulalumpur, pp. 298-310.

Choudhury, A., Bhunia, A.B. and Nandi, S. 1984b. Preliminary survey on macrobenthos of Prentice Island. Sunderbans, W.B. Rec. Zool. Surv. Ind. 81(3 & 4): 81-92.

Clark, R.B. 1964. Dynamics in Metazoan Evolution. Clarendon Press, Oxford, p. 313.

Colwin, A.L. and Colwin, L.H. 1949b. Development potencies of the egg of Saccoglossus (Dolichoglossus) kowalevskyii. Biol. Bull., 97: 237.

Colwin, A.L. and Colwin, L.H. 1949c. The fertilization reaction in the egg of Saccoglossus (Dolichoglossus) kowalevskyil. Biol. Bull., 97: 237.

Colwin, A.L. and Colwin, L.H. 1953. The normal Embryology of Saccoglossus kowalewsky. J. Morph., 92: 401-453.

Cook, D.G. and Johnson, M.G. 1974. Benthic macroinvertebrates of the St. Lawrence Great Lakes. J. Fish. Res. Bd. Can., 31: 763-782.

*Crozier, W. J. 1915. The behaviour of an Enteropneust. Science, 41: 471.

Crozier, W. J. 1917. The photic sensitivity of Balanoglossus. J. Exp. Zool., 24: 211.

*Damas, D. and Stiasny, G. 1961. Less larves planchtoniques d' enteropneustes (tornaria et planctosphaera). Verh. K. vlaam. Acad. Wet. (kl. Wetenschappen), 15 : 1-68.

Damodaran, R. 1973. Studies on the benthos of the Mud bank regions of the Kerala coast. Bull. Dept. Mar. Univ. Cochin, 6: 1-126.

Davis, B.M. 1980. The early life history of Dolichoglossus pusillus Ritter. Univ. Calif. Publ. Zoo. 4 : 187-226.

*Davis, D.S. 1967. The marine fauna of the black water estuary and adjacent waters. Essex Naturalist, 32 : 2-61.

*Dawydoff, C. 1928. Quelques observations sur le development des enteropneustes. C.R. Acad. Sci., Paris, T. 186 : 173.

Day, J.H. 1951. The ecology of South African estuaries Part I. A review of estuarine condition in general. Royal Soc. Afr. Transae 33: 53-91.

Day, J.H. 1981. Estuarine ecology, with particular reference to Southern Africa. Rotterdam : Balkema.

De Jorge, F.B.. Petersen, J.A. and Sawaya, P. 1967. Biochemical studies on the enteropneusta Balanoglossus gigas (Fr. Muller, 1898). Comp. Biochem. Physiol. 22: 467-475.

De Jorge, F.B. and Petersen, J.A. 1968a. Accumulation of "glycogens" in the genital region of Balanoglossus gigas Fr.

Muller (Spengel 1893) Enteropneusta during reproduction. Comp. Biochem. Physiol., 26 : 737-740.

De Jorge, F.B. and Petersen, J.A. 1968b. Comparative biochemical studies on Balanoglossus clavigerus and Willeyia loya (Enteropneusta). Comp. Biochem. Physiol., 27 : 551-557.

*Delle Chiaje, S. 1829. Memorie sulla storia e notomia degl animal senza vertebre del rengo di Napoli., 4:

Desai and Krishnanakutty, M. 1967. Studies on the benthic fauna of Cochin backwater. Proc. Ind. Acad. Sci. 66(4): 123-142.

Devanesan, D.W. and Varadarajan, S. 1940. The occurrence of Tornaria larva' at krusadai. Curr. Sci., 9 : 375-377.

Devanesan, D.W. and Chacko, P.I. 1942. Balanoglossus as food for fishes. Curr. Sci., 29: 242-243.

Dexter, D.M. 1979. Community structure and seasonal movements of subtidal benthic invertebrates in the Fraser river estuary. B.C. Estuar. Coast. Mar. Sci., 9: 543-558.

Diener, R.A. 1973. Observation on the hydrology and marine organism of the tidal colorado river and adjacent waters, Texas. June 1962. Contrib. Mar. Sci., 17: 99-110.

Driscoll, E. G. 1967. Attached epifauna substrate relations. Limnol. Oceanogr., 12: 633-640.

Dilly, P.N. 1975. The dormant buds of Rhabdopleura compacta (hemichordata). Cell Tissue Res. 159 : 387-397.

Dutta, A.K., Pal, P.K., Ghosh, A., Misra, S., Nandi, S. and Choudhury, A. 1986. Lipids and fatty acids of the gastropod mollusc Cerethidea cingulata. J. Am. Oil Chem. Soc. 63: 223-226.

Ellis, M.M. 1936, Erosion silt as a factor in aquatic environments. Ecology, 17: 29-42.

Elmhirst, R. 1934. Enteropneusta in the clyde sea Area. Nature Lond., 134: 183.

*Eschscholtz, F. 1825. Bericht uber die zoologische Ausbeute der Reise von kronstadt bis st. Peter and Paul. Oken's Isis for 1825, Heft, 6:

Evgen Eva, T.P. 1974. Cultivation of the notochord of Balanoglossus sp. in the diffusion chambers. Ontogenez, 5(1): 76-79.

Fernando, S.A., Khan, S.A. and Kashinathan, R. 1981. Observation on the distribution of benthic fauna of the Vellar estuary, Portonovo. Seminar on estuaries their Phy. Chem. Biol. Geo. and Engineering aspect. Abs. of Contrib. NIO, Goa, p. 13.

Ganapati, P.N. and Chandrasekhara Rao, G. 1962. Ecology of the intertidal fauna inhabiting the sandy beach of Waltair coast. J. Mar. Biol. Ass. Ind. 4: 44-57.

Ganapati, P.N. and Lakshmana Rao, M.V. 1962. Studies on the ecology of the intertidal sands of the Vishakhapatnam coast. Proc. 1st All India Congr. Zool. 1959, Pt. 3, Symposia. pp. 14-25.

Garstang, W. 1939. Spolia Bermudina. Part II. The ciliary feeding mechanism of Tornaria. Quart. J. Micro. Sci. 81 : 347-366.

Ghosh, A. and Dutta, J. 1972. Hydrogenator for micro-scale hydrogenation of fatty acids. Trans. Bose Res. Inst..35 : 13-16.

Ghosh, A., Misra, S. Dutta, A.K. and Choudhury, A. 1985. Pentacyclic triterpenoids and sterols from seven species of mangrove. Phytochemistry 24 : 1725-1727.

Ghosh, M. and Choudhury, A. 1986. Growth of Acanthamoeba culbertsoni (Singh and Das 1970) and Acanthamoeba rhysodes (Singh 1952) (Protozoa: Gymnamoebia) in cultures. Proc. Indian Acad. Sci. (Anim. Sci.), 95(4): 469-473.

Gilmour, T.H.J. 1979. Feeding in Pterobranch hemichordates and the evolution of gill slits. Can. J. Zool., 57(5): 1136-1142.

*Girard, C. 1853. New nemerteans from the coast of the carolinas. Proc. Acad. Natur. Sci., Philadelphia, 6.

Gunter, G. 1950. Seasonal population changes and distribution as related to salinity of certain invertebrates of the Texas coast, including the commercial shrimp. Publs. Inst. mar. Sci. Univ. Tex., 1(2): 7-51.

Gunter, G. 1956. Some relations of faunal distributions to salinity in estuarine waters. Ecology, 37: 616-619.

Gunter, G. 1961. Some relation of estuarine organisms to salinity. Limnol. Oceanogr., 6: 182-190.

Hadfield, M.G. 1975. 'Hemichordata'. In roproduction of Marine Invertebrates. Vol. II. Ectoprocts and lesser coelomates Giese, A.C. and Pearse, J.S. (Eds.). Academic Press, London, pp. 185-240.

Hadfield, M.G. and Young, R.E. 1983. Planctosphaera new record (Hemichordata : Enteropneusta) in the Pacific Ocean. Mar. Biol. (Berl.) 73(2): 151-154.

Hammer, S.F. 1904. Hemichordata. C.N.H., 7: 1-32.

Harkantra, S.N. 1975. On kali estuary of Karwar. Mahasagar Bull. natn. Inst. Oceanogr., 8 : 53-58.

Harkantra, S.N., Nair, A., Ansari, Z.A. and Parulekar, A.H. 1980. Benthos of the shelf region along west coast of India. Ind. J. mar. Sci., 9 : 106-110.

Harkantra, S.N., Rodrigues, C.L. and Parulekar, A.H. 1982. Macrobenthos of the shelf off north eastern Bay of Bengal. Ind. J. mar. Sci. 11 : 115-121.

Heider, K. 1909. Zur entwickleng von. Balanoglossus clavigerus (D.Ch.). Zool. Anz. 34: 695-704.

*Hess. W.N. 1936. Reaction to light in Ptychodera bahamensis. Pap. Tortugas Lab., 31 : 79.

Hess, W.N. 1937. The nervous system of Dolichoglossus kowalevskyi. J. Comp. Neurol., 68 : 161.

Hess, W.N. 1938. Reactions to light and the photoreceptors of Dolichoglossus kowalevskyi. J. Exp. Zool. 79: 1.

Hinrichs, V.O. and Jacobi, L. 1938. Saccoglossus pygmaeus eine noue Enteropneusta nartaus der sudlicheu Nordsee. Zool. Anz. Bol., 121: 25-32.

Holme, N.A. and McIntyre, A.D. 1971. Methods for the study of Marine Benthos, Oxford, U.K. : BlackwellScientific Publications, p. 334

Hymen, L.H. 1959. The Invertebrates. McGraw Hill Book Company, Inc., New York, 5 : 72-207.

*Ikeda, I. 1908. On the swimming habit of a Japanese enteropneust Glandiceps haeksii. Annot Zool. Jap., 6 : 255.

Jackson, M.L. 1973. Soil chemical Analysis. Prentice Hall of India Private Ltd., New Delhi.

Jumars Peter, A. and Robort, R. Hessler 1976. Ilada community structure : Implications from the Aleutian Trench. (for Enteropneusta). J. Mar. Res., 34(4): 547-560.

Kapelus, F.F. 1936. The anatomy of the Enteropneusta Saccoglossus inhacensis sp. n. Ann. Natal., 8 : 47-94.

Karriga, P. 1947. Relation between the moon and periodicity in the breeding of marine animals. Ecol. Monographs., 17 : 347-381.

Kemp, S. 1917. Notes on the fauna of the Matla river in the Gangetic delta. Rec. Ind. Mus. 13: 233-241.

Khan, S.A. and Natarajan, R. 1975. Ecology of hermit crab population in vellar estuary at Portonovo south India. In recent researches in estuarine biology. R. Natarajan (Ed.). H.P.C. (I) Ltd., NDLH., pp. 255-272.

Kinne, O. 1963. The effect of temperature and salinity on marine and brackish water animals. Oceanogogr. Mar. Biol. Ann. Rev. 1: 301-340.

Kinne. O. 1964. The effect of temperature and salinity on marine and brackish water animals. Oceanogr. mar. Biol. Ann. Rev., 2 : 281-339.

Kirk, H.B. 1937. The embryology of Dolichoglossus otagoensis Benham (Abstract). Rep. Anat. Assoc. Adv. Sci..23 : 137-138.

Kirk, H.B. 1938. Notes on the breeding habits and early development of Dolichoglossus otagoensis Benham. Trans. Proc. Roy. Soc., N.Z., 88: 49-50.

Knight-Jones, E.W. 1952. On the nervous system of Saccoglossus combrensis (Enteropneusta). Phil. Trans. Roy. Soc., London, B 236: 315-354.

Knight-Jones, E.W. 1953. Feeding in Saccoglossus (Enteropneusta). Proc. Zool. Soc. Lond., 123 : 637-691.

Knight-Jones, E. W. 1954. Relation between metachronism and the direction of ciliary beat in metazoa. Quart. J. micr. Sci., 95: 503-521.

Knight-Jones, E.W. and Millar, R.H. 1947. Bilateral asymmetry shown by the Metachronal waves in Protochordate gill slits. Nature Lond., 163: 137-138.

Komai, T. 1951. The homology of the 'Natochord' found in pterobranchs and Enteropneusta. Amer. Nat., 85: 270.

*Kowalevsky, A. 1866. Anatomie des Balanoglossus Delle chiaje. Mem. Acad. Imp. Sci. St. Peters Ser. 7 T.X. 3 : 18.

Krishnamurthy, B. 1963. On the distribution of six species of polychaetes in the Adyar estuary Madras. J. Mar. Biol. Ass. Ind..5: 57-102.

Krishnan, G. and Govindarajulu, P. 1968. Observations on the biochemical composition of the enteropneusta Ptychodera flava. Life Sci., 7 : 629-632.

*Krohn, A. 1854. Beobachtungen uber Echinodermen larven. Arch. Anat. Physiol. Wissensch. Med.

Kurian, C.V. 1967. Studies on the benthos of the south west coast of India. Bull. Nat. Inst. Sci. India 38: 649-656.

Kurian, C.V. 1967. Studies on the benthos of the mud bank of the Kerala coast. Bull. Dept. Mar. Sci. Univ. Cochin, 6: 1-126.

Kurian, C.V. 1971. Bottom fauna of a tropical estuary. Proc. Joint. OC. Assembly Tokyo. 'The Ocean World', pp. 455-457.

Kurian, C.V. 1972. Ecology of benthos in a tropical estuary. Indian natn. Sci. Acad., 38 (3 & 4) : 156-163.

Kuriyan, G.K. 1949. On the occurrence of Enteropneusta in Shingle Island, Gulf of Manaar. Curr. Sci., 18 : 258.

Lindeman, R. L. 1942. The trophic-dynamic aspect of ecology. Ecol. 23 : 399-418.

Macbride, E.D. 1894. Review of the Spengel's monograph of Balanoglossus. Quart. J. Micro. Sci., 36: 385-420.

Macha, N. and Petersen, J.A. 1969. A histochemical study of epidermal mucous secretions in Enteropneusta (Hemichordata). Acta Histochem., 32: 305-309.

*Maser, O. 1913. Balanoglossus carnosus. Zool. Jahrb. Anat. Abt., 33:

McLachlan, A. and Erasmus, T. 1983. Sandy beaches as ecosystems. The Hogue: Junk.

McNulty, J.K., Work, R.C. and Moore, H.B. 1962. Some relationships between the infauna of the level bottom and the sediment in South Florida. Bull. mar. Sci. Gulf. Caribb., 122: 322-332.

McIntyre, A.D. 1964. Meiobenthos of sub-littoral muds. J. mar. biol. Ass. U.K., 44: 665-674.

McIntyre, A.D. 1968. The meiofauna and macrofauna of some tropical beaches. J. Zool. Lond. 156: 377-392.

McIntyre. A.D. 1969. Ecology of marine meiobenthos. Biol. Rev. 44: 245-290.

Meek, A. 1922. Glossobalanus marginatus a new sp. of enteropneusta from the North sea. Quart. J. Micr. Sci., 66 : 579-594.

Melbourne, R.C. 1967. Ecology of estuarine benthic invertebrates a perspective estuary. Publns. Am. Ass. Advant. Sci., 83: 442-487.

Menon, K.R. 1904. Enteropneusta from Madras. Quart. J. Micro. Sci., 47: 123-131.

*Metschnikoff, E. 1869. Uber Tornaria. Nachtrichten Gesellsch. Wissensch. Univ. Gottingen. No. 15.

Misra, S., Dutta, A.K.. Dhar, T., Ghosh, A., Choudhury, A. and Dutta, J. 1983. Fatty acids of the mud skipper Boleophthalmus boddaerti. J. Sci. Food. Agric. 34 : 1413-1418.

Misra, S., Ghosh, M.K., Choudhury, A., Dutta, A.K., Pal, P.K. and Ghosh, A. 1985. Fatty acids from Macoma sp. bivalve mollusc. J. Sci. Food. Agri., 36: 1193-1196.

Morgan, T.H. 1891. Growth and metamorphosis of Tornaria. J. Morphol. 5: 407-458.

Morgan, T.H. 1894. Development of Balanoglossus. J. Morphol., 9: 1-86.

Morton, J.E. 1960. The functions of the gut in ciliary feeders. Biol. Rev., 35 : 95-140.

*Muller, J.E. 1850. Uber die larven and die Metamorphose der Echinodermen. Abhandl. Adad. Wissensch. Berlin (1848).

*Muller, Arno Hermann 1975. The animal macroplankton (Graptoloidea) of the silurian seas and some general data on Graptolithina (Hemichordata). Biol. Rundsch..13(6): 325-344.

*Muus, B.J. 1967. The fauna of Danis estuaries and lagoons : distribution and ecology of dominating species in the shallow water of the mesohaline zone Medd. Danm. Fisk-og Havus ders. N.S., 5 : 1-316.

Nagabhushanam, A.K. 1972. Studies on the marine intertidal ecology of Orissa coast. Proc. Indian Natn. Sci. Acad., 38(3 & 4) : 308-315.

Nagabhushanam, A.K. and Chandrasekhara Rao, G. 1969. A preliminary observations on a collection of shore fauna of the Orissa coast. Proc. Zool. Soc., Calcutta, 22 : 67-82.

Nandi, S. and Choudhury, A. 1983. Quantitative studies on the benthic macrofauna of Sagar Island, intertidal zones, Sunderbans, India. Mahasagar, Bull. natn. Inst. Oceanogr..16(3) : 409-414.

Nandi, S. 1986. Ecology of benthic gastropods in the littoral mudflats of Hooghly estuary, India. Doctoral Dissertation, Calcutta University.

Newell, G.E. 1949. Occurrence of Saccoglossus kowalevskyi on the North kent coast. Nature, 164 : 747.

Newell, G.E. 1952. The stomochord of Enteropneusta. Proc. Zool. Soc., London, 121 : 741-746

Newell, G.E. 1952. The Homology of the stomochord of the Enteropneusta. Proc. Zool. Soc., London, 121 : 741-746.

Norrevang, A. 1964. Choanocytes in the skin of Harrimania kupfferi (Enteropneusta). Nature, Lond., 204 : 398-399.

Norrevang, A. 1965. On the mucous secretion from the proboscis in Harrimania kupfferi (Enteropneusta). Ann. N.Y. Acad. Sci., 118: 1053-1069.

Odum, Eugene, P. 1971, Fundamentals of ecology W.B. Saunders Co. (Third ed.), p. 352.

Parulekar, A.H., Dhargalkar, V.K. and Singbal, S.Y.S. 1980. Benthic studies in Goa estuaries: Part III Annual cycle of macrofaunal distribution, production and trophic relations. Ind. J. mar. Sci., 9 : 189-200.

Parulekar, A. H. and Dwivedi, S.N. 1973. Ecology of benthic production during south-west monsoon. In recent researches in estuarine biology, R. Natarajan (Ed.). H.P.C. (I) Ltd. NDLH. pp. 21-30.

Parulekar, A.H. and Dwivedi, S.N. 1974. Benthic studies in Goa estuaries Part I. Standing crop and faunal composition in relation to bottom salinity distribution and substratum characteristics in the estuary of Mandovi river. Ind. J. Mar. Sci. 3: 41-45.

Parulekar, A.H., Nair, S.A., Harkantra, S.N. and Ansari, Z.A. 1976. Some quantitative studies on the benthos off Bombay. Mahasagar-Bull. Natn. Inst. Oceanogr., 9 : 51-56.

Parulekar, A.H. and Wagh, A. B. 1975. Quantitative studies on benthic macrofauna of north eastern Arabean sea shelf. Ind. J. mar. Sci. 4 : 174-176.

Paterson, C.G. and Fernando, C.H. 1971. Studies on the spatial heterogeneity of shallow water benthos with particular reference to the Chironomidae. Can. J. Zool. 49: 1013-1019.

*Payne, F. 1936. Early development of Ptychodera bahamensis. Pap. Tortugas Lab., 31 : 71-76.

Perkins, E.J. 1974. The biology of estuaries and coastal waters. London: Academic Press.

*Petersen, C.G.J. and Boysen Jensen, 1911. A valuation of the sea. I. Animal life of the sea bottom its food and quantity. Rep. Danish. Biol. Stn. 20: 3-79.

*Petersen, J.A. and Longhi, L. 1971. Amino acid distribution in Glossobalanus (Ptychoderidae, Enteropneusta, Hemichordata). Histochemistry, 16 : 323-327.

Pillay, T.V.R. 1958. Hilsa ilisha (Ham.) of the river Hooghly. Indian J. Fish. 5: 1.

Piper, C.S. 1950. Soil and plant analysis. Academic Press, New York.

Pritchard, D.W. 1967. What is an estuary? Physical viewpoint. In: Estuaries. G.H. Lauff (Ed.). Amer. Assoc. Adv. Sci. Publ. Washington DC, 83 : 3-5.

Punnett, R.C. 1903. The enteropneusta in : The fauna and geography of the Maldive and Laccadive Archipelagoes. J.S. Gardiner (Ed.). 2 : Part Cambridge.

Purchon, R.D. and Jefferson, G.T. 1949. A hemichordate from west wales. Nature, London, 164 : 881.

Ramanujam, S.G.M. 1935. Enteropneusta from Krusadai Island. Curr. Sci. 3: 427.

Rainer, S.F. 1981. Soft bottom benthic communities in Otago Harbour and Blueskin Bay, Newzealand. N.Z. Oceanogr. Inst. Meno, 80: 5-38.

Rao, K. Pampapathi, 1951. The anatomy of Ptychodera flava and studies on other Enteropneusta from Madras. Doctoral Dissertation, Univ. of Madras.

Rao, K. Pampapathi 1952. Significance of variation in P. flava, Evolution, 6: 342-343.

Rao, K. Pampapathi 1953. Development of Glandiceps. Jour. Morphol..93: 1-8.

Rao, K. Pampapathi 1954. Bionomics of Ptychodera flava. Jour. Madras Univ. Sect. B. 24(1-3): 1-4.

Rao. K. Pampapathi 1955a. Early development of Ptychodera flava. Jour. Zool. Soc., India, 6(2): 145-152

Rao, K. Pampapathi 1955b. Enteropneusta from Madras and the gulf of Manaar, with a note on the probable course of distribution of Ptychodera flava Eschscholtz. Proc. Ind. Sci. Cong. 42 Pt. III, Sect 7: 301.

Rao, K. Pampapathi 1955c. Two new species of Enteropneusta from Madras. Proc. Ind. Sci. Cong.. 42. PL. III, Sect. 7: 301.

Rao, K. Pampapathi 1955d. Tornaria from Madras. Hydrobiologia, 7: 269-278.

Rao, K. Pampapathi 1955e. Morphogenesis during regeneration in an Enteropneusta. Jour. Animal. Morphol. Physiol., 1(1): 1-7.

Raymond, B.A. and Milton, J.C. 1967. Isolation of 2, 6-Dibromophenol from the marine Hemichordates Balanoglossus biminiensis. Science, 155 : 1558-1559.

Rhoads, D.C. and Young, D.K. 1970. The influence of deposit feeding organisms on sediment stability and community trophic structure. Journal of Marine Research, 28(2): 150-178.

*Ritter, W.E. 1902. The movements of the Enteropneusta and the mechanism by which they are accomplished. Biol. Bull. (Wood Hole), 3: 255.

*Ritter, W.E. and Davis, B.M. 1904. Studies on the ecology morphology and specialogy of the young of some Enteropneusta of western North America. Univ. Calif. Publ. Zool., 1: 171-210.

Robinson, V.C. 1927. Report on a new species of Enteropneusta. Trans. Zool. Soc., London, 22: 361-364.

Sanders, H.L. 1956. The biology of marine bottom communities. Bull. Bingham Ocean. Coll., 15 : 345-415.

Sanders, H.L. 1958. Benthic studies in Buyzards Bay I. Animal sediment relationship. Limn and Ocean 3 : 245-258.

Sanders, H.L. 1960. Benthic studies in Buzyards Bay. III. The structure of the softbottom community. Limn. Ocean., 5: 138-153.

Segerstrale, S.G. 1978. The negative correlation between the abundances of the amphipod Pontoporeia and the bivalve Macoma in Baltic waters, and the factors involved. Arn. Zool. Fenn., 15 : 143-145.

Sephton, T.W. and Fernando, C.H. 1980. Spatial interrelationships of bivalves and nonbivalve benthos in a small reservoir in New Brunswick, Canada, Can. J. Zool., 58(5): 852-859.

Seshappa, G. 1953. Observation on the physical and biological features and the inshore sea bottom along the Malabar coast. Proc. natn. Inst. Sci. Ind., 19 : 257-279.

Seshappa, G. 1956. Observations on the composition of bottom muds in relation to phosphate cycle in the inshore waters off Malabar coast. Proc. Indian Acad. Sci., B 43 : 288-301.

*Schepotieff, A. 1908. Knospungs Prozess von Cephalodiscus. Zool. Jahrb. Abt. Anat., 25.

*Schimkewitsch, W. 1882. Uber die Beziehung Zwischen den Enteropneusta and Acrania. Revue Sci. Natur. St. Petersbourg (Viestnik Estesvoznaniia), 3.

*Schimkewitsch, W. 1888. Uber Balanoglossus mereschkowskii Wagner. Zool. Anz., 11: 250.

Schlenk, H. and Gallerman, J.L. 1960. Esterification of fatty acids with diazomethane on a small scale. Anal. Chem., 32: 1412.

*Silen, L. 1950. On the nervous system of Glossobalanus marginalis Meek (Enteropneusta). Acta. Zool. Stockh., 31 : 140.

Silen, L. 1954. Reflection concerning the Stomochord. Proc. Zool. Soc. Lond., 124 : 63.

Singh, B. and Choudhury, A. 1984. Occurrence of an Enteropneust Hemichordate in the Mangrove swamp of Sundarbans, India. Bull. Zool. Surv. India, 6(1-3): 1-4.

Singh, B.N. and Choudhury, A. 1985. Morphological excellence, feeding and breeding behaviour of Saccoglossus sp. (Hemichordata, Enteropneusta) from mangrove mudflats of Sundarbans, India. The Mangroves, Proc. Nat. Symp. Biol. Util. Cons.. Mangroves pp. 505-510.

Snedecor, G.W. and Cochran, W.G. 1967. Statistical methods. The Lowa State Univ. Press, U.S.A.

*Spengel, J.W. 1891. Uber die Gattungen der Enteropneusten. Verhandl. Dtsch. Zool. Gesellsch, 1.

*Spengel, J.W. 1893. Die Enteropneusten des Golfes von Neapel. Fauna and Flora des Golfes von Monogr, 18.

*Spengel, J.W. 1901. Die Benennung der Enteropneusten-Gattunge Zool. Jahrb. Abt. System., 15.

*Spengel, J.W. 1904. New Beitrage zur kenntnis der Enteropneusta. Zool. Jahrb. Abt. Anat., 20.

Spengel, J.W. 1907. Studies uber dies Enteropneustan der siboga Expedition. Siboga Exped Monogr., 26.

Spengel, J.W. 1909. Pelagisches voskommen vou Enteropneusta. Zool. Anz., 34.

Stephen, A.C. and Edmonds, S.J. 1972. The phyla siphuncula and echiura. Trustees of the British Museum (Nat. Hist.), London, England, 10.

*Stiasny, G. 1910. Zur kenntniss der lebensweise von Balanoglossus clavigerus. Zool. Anz., 35 : 561.

*Stiasny. G. 1912. Die sandwirben (Faceshaufchen) des Balanoglossus clavigerus. Zool. Anz., 39 : 488.

*Stiasny, G. 1913. Studies uber die Entwicklung von Balanoglossus clavigerus. D. Ch. Zool. Bd., pp. 487-500.

Stiasny-Wijnhoff, G. and Stiasny, G. 1926. Uber Tornarian typen and ihre Beziehung zur systematic der Enteropneustan. Zool. Anz., 68: 159-165.

Strathmann, R. and Bonar, D. 1976. Ciliary feeding of Tornaria larva of Ptychodera flava (Hemichordata; Enteropneusta). Mar. Biol. (BERL.), 34(4): 317-324.

Strickland, J.D.H. and Parsons, T.R. 1968. A manual of seawater analysis. Fish. Res. Bd. Canada, Bull. 167 : 311.

Subbiah, B.V. and Asija, G.L. 1965. A rapid procedure for estimation of available nitrogen in soils. Curr. Sci. 25(8): 259.

Sundara Rao, P.J. and Rangarao, S. 1949. A note on the occurrence of a giant Balanoglossid at Krusadai Island. J. Bomb. Natu. Hist. Soc. 48 : 813.

Tattersall, W.M. 1905. Enteropneusta from the west coast of Ireland. Ann. Rep. Fish. Ireland Sci. Invest., 1902 : 213-214.

Tenore, K.R. 1972. Macrobenthos of the Palmico river estuary. North Carolina. Ecol. Monogr., 42 : 51-69.

Thomas, I.M. 1955. Saccoglossus apantesis a new species of enteropneust from South Australia. Trans. R. Soc. S. Aust., 79 : 167-176.

Trueman, E.R. and Ansell, A.D. 1969. The mechanisms of Burrowing into soft substrata by Marine animals. Occeanogr. Mar. Biol. Ann. Rev.,7 : 315-366.

Turpaeva, E.P. 1959. Food interrelationship of dominant species in marine benthic biocoenoses. In Marine Biology, pp. 137-148. Nikitin, B.N. (Ed.). Trans. Inst. Oceanology, 20.

Van der Horst, C.J. 1932. Die Enteropneusten Gattung Ptychodera. Zool. Anz. Ecol., 99 : 98-100.

*Van der Horst, C.J. 1932. Enteropneusta In Handbuch der Zoologie (Kukenthal and Krumbach), 3 2(9) : 1-56.

Van der Horst, C.J. 1934b. The burrow of an Enteropneusta. Nature, 134 : 852.

Van der Horst, C.J. 1935. Planctosphaera and Tornaria. Quart. Jour. Micr. Sci., 78 : 605.

*Van der Horst, C.J. 1939. Hemichordata. In H.G. Bronn (Ed.). Klassen and Ordnungan des Tierreichs, 4 : Abt. Buch 2, Teil 2.

*Van der Horst, C.J. 1940. The enteropneusta from Inyack Island, Delagoa Bay. Ann. S. Afr. Mur., 32 : 293.

Varshney, P.K., Nari, A., Ansari, Z.A. and Parulekar, A.HH. 1981. Benthos of the shelf region along the West coast of India. Ind. J. mar. Sci., 9 : 106-110.

*Wiley, A. 1899. Enteropneusta from the south pacific with notes on the west Indian species. Wiley's Zoological Results, 3: 223-234.

Young, D.K. 1968. Effect of infauna on the sediment and seston of a subtidal environment. Abstr. Third Europ. Mar. Biol. Symp., Aracachon, France.

Young, D.K. and Rhoads, D.C. 1971. Animal-sediment relations in Cap Cod Bay, Massachusetts. I. A transect study. Marine Biology 11 : 242-254.

***Not seen in original.**